International Marketing in the Network Economy

'This book is a consequent and smart application of the knowledge creation theory to the field of marketing. The proposed shift in marketing towards the knowledge-based paradigm has vast theoretical and practical implications. Kohlbacher convincingly illustrates how companies create and leverage marketing knowledge to successfully compete in the network economy of today.' – Hirotaka Takeuchi, Professor and Dean, Graduate School of International Corporate Strategy, Hitotsubashi University, Tokyo; co-author of *The Knowledge-Creating Company: How Japanese Companies Create the Dynamics of Innovation*.

'Kohlbacher shows a formidable ability to integrate state-of-the-art theory with data from world-leading companies. His treatment of knowledge as the basis for marketing success is truly international. He assumes the dual roles for reflective scholar and reflective practitioner, not least by respecting the need for both systematic marketing research and experiential and tacit knowing. It is not only a seminal contribution to research but a platform for the rejuvenation of executive training.' – Evert Gummesson, Professor, Stockholm University School of Business, Sweden; author of *Total Relationship Marketing: From 4Ps to 30Rs*.

'Dr Kohlbacher synthesizes a vast amount of research to create a convincing argument that the canny marketing manager has an impressive array of knowledge-based capabilities available for development and leveraging – especially the tacit knowledge that is so critical to lasting advantage. The book is a welcome, significant addition to the growing body of academic research on managing knowledge assets.' – Dorothy Leonard, William J. Abernathy Professor of Business, Emerita, Harvard Business School; author of *Wellsprings of Knowledge: Building and Sustaining the Sources of Innovation*.

'Knowledge management approaches and techniques have mostly found their application in functions such as R&D, manufacturing, logistics, and service. This book explores the value of knowledge in marketing and how the creation of so called "marketing knowledge" can be more effectively enabled. The book is a must-read for anyone interested in the creation and transfer of marketing knowledge. The many case studies in the book provide great insights for managers who strive to improve marketing in their firms.' – Georg von Krogh, Professor and Chair of Strategic Management and Innovation, ETH Zurich; co-author of *Enabling Knowledge Creation: How to Unlock the Mystery of Tacit Knowledge and Release the Power of Innovation*.

'The marriage of marketing and knowledge management – and specifically knowledge-creation theory – is long overdue. Kohlbacher's timely book challenges the academic marketing community to rethink its assumptions about the nature of marketing knowledge as a competitive resource especially in international and cross-cultural contexts.' – Nigel J. Holden, Professor and Director, Institute of International Business, Department of Strategy and Innovation, Lancashire Business School, University of Central Lancashire; author of *Cross-Cultural Management: a Knowledge Management Perspective*.

International Marketing in the Network Economy

A Knowledge-Based Approach

Florian Kohlbacher
Research Fellow, German Institute for Japanese Studies, Tokyo, Japan

Foreword by Ikujiro Nonaka
and
Afterword by Charles M. Savage

First published 2007 by
PALGRAVE MACMILLAN
Houndmills, Basingstoke, Hampshire RG21 6XS and
175 Fifth Avenue, New York, N.Y. 10010
Companies and representatives throughout the world

PALGRAVE MACMILLAN is the global academic imprint of the Palgrave
Macmillan division of St. Martin's Press, LLC and of Palgrave Macmillan Ltd.
Macmillan® is a registered trademark in the United States, United Kingdom
and other countries. Palgrave is a registered trademark in the European
Union and other countries.

ISBN 13: 978–0–230–51570–3   hardback

This book is printed on paper suitable for recycling and made from fully
managed and sustained forest sources. Logging, pulping and manufacturing
processes are expected to conform to the environmental regulations of the
country of origin.

A catalogue record for this book is available from the British Library.

Library of Congress Cataloging-in-Publication Data

Kohlbacher, Florian.
    International marketing in the network economy : a knowledge-based
    approach / Florian Kohlbacher; foreword by Ikujiro Nonaka and
    afterword by Charles M. Savage.
        p. cm.
    Includes bibliographical references and index.
    ISBN 978-0-230-51570-3 (alk. paper)
    1. Export marketing. 2. Knowledge management. I. Title

    HF1416.K64 2007
    658.8'4–dc22   2007023305

Contents

List of Figures

Foreword

Dr Ikujiro Nonaka

In this book, Florian Kohlbacher proposes a knowledge-based theory of marketing that is the first of its kind and a significant contribution to the fields of both marketing and knowledge management.

My own roots are in marketing. I was interested originally in information processing and spent five-and-a-half years at the University of California–Berkeley, completing my PhD dissertation in 1972 (Nonaka, 1972) majoring in marketing. Francisco M. Nicosia was my mentor and I studied consumer decision processes under his guidance (see, for example, Nonaka and Nicosia, 1979). His work at the time was based on behavioural science and his major contribution was the conceptualization of consumer decision processes from the perspective of information processing, so it was fitting that Herbert Simon wrote the preface to his book.

Under Nicosia's influence, my interest shifted from marketing to organization theory and I became interested in the process of knowledge creation (Nonaka, 2005). The turning point came when I participated with my colleagues Hirotaka Takeuchi and Ken-ichi Imai in the 75th Anniversary Colloquium on productivity and technology at Harvard Business School. We agreed then to do a joint study of the innovation processes at several Japanese companies. We presented the results in a paper entitled, 'Managing the New Product Development Process: How Japanese Companies Learn and Unlearn'. I found that innovation is not simply about information processing, but is also a process of capturing, creating, leveraging, and retaining knowledge.

The framework presented in this book shows that management of new product development is chiefly a marketing process where knowledge and knowledge creation play critical roles. It points out that marketing is one of the most knowledge-intensive activities of a company. My own theory of organizational knowledge creation and Dorothy Leonard's work on the wellsprings of knowledge (Leonard, 1998) focus on new product development and innovation as a key marketing process. Both theories are built around examples of knowledge-based marketing, but we focused primarily on the process, and on developing a model of knowledge creation. Finally, this book synthesizes our original ideas with more recent insights in a comprehensive model of knowledge-based marketing.

I sincerely welcome the new and fresh approach to marketing it proposes. Academic marketing has never been able to break free of its roots in neoclassical economics, and practical marketing is often regarded as merely one business function among others in a firm. The research stream has remained stuck in information processing theory and models of learning.

As this book shows, marketing is about much more than passive learning and information processing. It is about knowledge creation and co-creation. The business ecosystem is a reservoir of knowledge that can be leveraged through human interaction. The human being and human activity are at the centre. Indeed, the most important entity in the business ecosystem is still the human being with her rich tacit knowledge and deep relationships.

For a knowledge-creating firm, environment is not an abstract object of scientific analysis but a phenomenological 'life-world' that is lived and experienced (Husserl, 1954). Rather than examining the environment objectively as industrial structure, managers are thrown into strategic decision-making as a way of life. The employees at Maekawa described in Chapter 5 are encouraged to think *as* customers instead of thinking *for* them. Their preconceived notions will prevent them from seeing customers as they are, if they are viewed merely as subjects of scientific analysis. But the phenomenological method of 'seeing the environment as it is' does not imply unconditional acceptance of it. By pursuing an understanding of the essence of environment, interpreted through dialogue and practice, knowledge arises from the interpretations (Nonaka and Toyama, 2005).

The ecosystem of knowledge consists of multilayered *ba*, which exist across organizational boundaries and are continuously evolving. Firms create knowledge by synthesizing their own knowledge with the knowledge embedded in various outside players such as customers, suppliers, competitors, and universities. Through interaction with the ecosystem, a firm creates knowledge, and that knowledge changes the ecosystem. Organization and environment should thus be understood as evolving together rather than as separate entities. The continuous accumulation and processing of knowledge help firms to redefine their vision, dialogue, and practice, which, in turn, impacts on the environment through new or improved services or products (Nonaka and Toyama, 2006b).

This dynamic relationship is difficult to grasp from the traditional viewpoint of markets and organizations that is inherent in organizational economics. Firms in the ecosystem of knowledge can no longer define their existence on the basis of ownership. Boundary-setting based on transaction cost is insufficient for understanding and managing competitive advantage based on knowledge. A knowledge-creating firm must be able to manage multi-layered *ba*, which stretch beyond organizational boundaries. At the same time, the firm has to protect its knowledge assets as a source of competitive advantage. Viewed in this context, the protection of knowledge assets is a complex and arguably impossible task (Nonaka and Toyama, 2005b).

Organizational knowledge creation is a dynamic process starting at the individual level and expanding through communities of interaction that transcend sectional, departmental, divisional, and even organizational boundaries. Firms acquire and take advantage of the tacit knowledge embedded in customers and suppliers by interacting with them. Organizational knowledge creation is a never-ending process that continuously upgrades itself. This interactive spiral exists both intra- and inter-organizationally. Knowledge transferred beyond organizational boundaries is synthesized with knowledge from different organizations to create new knowledge. In this dynamic interaction, knowledge created by the organization triggers a mobilization of knowledge held by outside constituents such as consumers, distributors, affiliated companies, and universities. It enables communication of the tacit knowledge possessed by customers that they themselves have not been able to articulate. An innovative manufacturing process may bring about changes in supplier manufacturing processes, which, in turn, trigger a new round of product and process innovation in the organization. A product works as a trigger to elicit the tacit knowledge derived from customers giving meaning to the product by purchasing, using, adapting, or not purchasing it. Their actions are then reflected in the innovation process of the organization, launching a new spiral of organizational knowledge creation.

In my previous work, I have focused primarily on the organizational knowledge-creating process within a company. I have described the process as a dynamic interaction between organizational members, and between organizational members and the environment. But the knowledge-creating process is not confined within the boundaries of a single company. The market is also a place for knowledge creation as it is where the knowledge of companies and consumers interacts. It is also possible for groups of companies to create knowledge. If we raise the level of analysis further, we can discuss how so-called national systems of innovation can be built. Therefore, it is important to examine how companies, governments, and universities can work together to make knowledge creation possible (Nonaka, Toyama, and Konno, 2000).

Returning to the concept of knowledge-based marketing, it is striking that even when talking about customer-focus and customer-centric firms, both marketing scholars and practitioners fail to make the conceptual leap necessary to overcome the separation of subject and object that makes discussion of these possible. What is needed is the holistic view of marketing or knowledge-based marketing proposed in this book. Marketing is not only about listening to and knowing the customer. Firms and managers have to take the standpoint of the customer, and collaborate with them and dwell in them to share and co-create tacit knowledge. This can help the firm to grasp customers' latent needs – needs that customers themselves are yet unaware of. The role of *ba*, which involves deep listening to and empathizing with customers and other entities in the business ecosystem, is crucial. Moreover, if we just stick to conventional marketing 'techniques' of presentation or knowledge models such as CRM (customer relationship management), we miss the qualitative aspects of judgement that enliven these approaches.

In that respect, this book highlights in a 'qualitative' research approach some very important lessons that might never show up in the traditional 'quantitative' method. Of course, we need both, and yet some still believe that without a strictly quantitative approach the results are not valid. That is certainly not the case with the systematic approach to the selection of informant companies and case studies found here.

Let me emphasize once again the importance of *ba*. A theory of knowledge-based marketing must also be one of contextual marketing. Looking at the case studies presented in this book, there are many examples of '*ba*-creating marketing', such as the communities of practice at Hewlett Packard Consulting & Integration Japan, or the introduction of the new Schindler escalator in Asia, as well as the co-creation of *ba* and empathy at Mazda and Maekawa. The transcendence of one's subjective view towards in-dwelling with the customer is crucial. Knowledge-based marketing must also be social marketing. Siemens One is probably the most comprehensive and systematic case of knowledge-based marketing. But here also, *ba* plays an important role as the shared context that has to be established to bridge the boundaries between different business units and departments to enable effective collaboration and knowledge creation. In the case of Toyota's joint venture with PSA, this might be more difficult as *ba* has to be co-created with a competitor with a very different national and corporate culture.

A shift in marketing thinking and theory towards the knowledge-based paradigm also has vast practical implications. As perspectives and everyday activities change, marketing practice will also change. Knowledge-based marketing and co-creation of marketing knowledge will have to be found in the relationships between people and between the firm and other entities in the business ecosystem. The contribution of this book to the theory of knowledge-based marketing is particularly valuable in this respect.

As a fine blend of marketing and knowledge-based management, the research skilfully integrates theory and practice. Based on a thorough and comprehensive review of existing research in the fields of both knowledge management and marketing, it proposes a theory of knowledge-based marketing supported by extensive empirical analysis. The case studies sampled are vivid examples of companies that have consciously taken a knowledge-based approach to marketing. Analysis and discussion of these elicit essential conclusions for both academics and practitioners.

Finally, the book argues convincingly that the view of marketing as just another corporate function will have to change, to one that sees it as part of the process of strategic management. Thus, the theory of knowledge-based marketing is also one of strategic management.

As such, this book is an important milestone in building a knowledge-based theory of the firm, and provides direction for further research based on its comprehensive and systematic framework. But real change in the mindset of marketing scholars and practitioners will take a lot more time and effort. I therefore encourage the continued pursuit of this path in research by the author and his readers towards co-creating the future of marketing and management in the knowledge economy of the twenty-first century.

Professor Emeritus
The Graduate School of International
Corporate Strategy,
Hitotsubashi University, Tokyo
Xerox Faculty Fellow in Knowledge, IMIO
University of California, Berkeley
The First Distinguished Drucker Scholar
in Residence at the Drucker School
and the Drucker Institute, Claremont
Graduate University

Preface

So this is yet another book on knowledge management? How many more treatments of this issue do we need, as there are already thousands of books, articles, theses, and so on, around? What is the new aspect, the value-added of this particular work? It is questions like these that I have come across frequently ever since my interest in so-called knowledge management began in 2002 and especially since I started my own research project in 2004. And I have been – and still am – asking myself these questions. In answer to the first question: No, this is actually not a book on knowledge management, at least not about knowledge management in the traditional sense of the term. In the course of reviewing hundreds of articles and books on the topic as well as when interviewing and talking with hundreds of people – among them many experts in the field – I realized that the term 'knowledge management' can easily cause misconceptions and misunderstandings. This has to do with the way the term has been used by certain people, especially IT experts and consultants, but also with misconceptions about 'managing' knowledge. I have therefore tried to find a different term that better explains what my research is about and have struggled for a long time to find an appropriate one. This was indeed a difficult task since my research is about knowledge as well as about management, but it is not necessarily about managing knowledge itself. Rather, it is about creating, sharing, transferring – recreating – knowledge and – very importantly – about applying and using this essential resource. It is about managing organizations and tasks on the basis of knowledge. Therefore, I decided to use the term *knowledge-based management* or *knowledge-based approach to management* to stress this important difference.[1] I am still not quite sure if this is the perfect term for it since it still looks similar to knowledge management, but at least it conveys one important notion and it is somewhat different from the traditional term. Nevertheless, while I strongly argue that firms should not be concerned with knowledge management but with knowledge-based management, I still tend to use the term 'knowledge management' to denote the academic field concerned with issues of knowledge, its management and knowledge-based management in firms.

The second question is whether there is still a need for more research in the field. Having written up my research I am convinced that the answer is yes. Many academics and practitioners still seem not to have grasped the importance of knowledge – especially the tacit part of it – as a source of competitive advantage and many still seem to hang on to certain misconceptions and misunderstandings about managing knowledge. This is why I hope this book will offer some new insights and value-added by unveiling some of these misconceptions and by establishing knowledge-based approaches to marketing, thus proclaiming a new dominant logic towards which firms have to evolve. But in the end, it will be up to the readers to decide whether there is any value-added or not.

Basically, the questions I mentioned above can be summarized and synthesized into one concise and extremely powerful question: 'So what?' Gabriel Szulanski relates a short anecdote in his book on barriers to knowing in the firm when his thesis adviser – looking at Szulanski's dissertation – asked him pointedly: '"So what? What should I do differently as a researcher because of your findings? What should managers do differently because of your findings? So what?"' (Szulanski, 2003: viii).

After more than two years of research and copious pages of manuscript writing, this question is a key challenge also to my research project. This book introduces knowledge-based management concepts to the field of marketing and presents a conceptual framework of knowledge-based marketing. Since this is a novel approach in marketing and management science, it provides an important academic contribution. So what? The crux is what will happen from now on. How will this new theory be received by the scientific community? Will it be noticed at all? And if yes, will it have an impact and of what kind? I believe the framework provides a powerful tool for analysing marketing processes from a knowledge-based perspective, but it will only be in its application that an answer to the 'so what?' question will emerge. The same is true for its practical business contribution. The framework can help managers to grasp the importance of knowledge in marketing and how to leverage the power of marketing knowledge co-creation both within the firm and within other entities in the firm's business ecosystem. So what? Only when managers and firms start to apply the concepts presented here and really venture such a knowledge-based approach to marketing will we discover the answer. But I take heart from the examples of the companies presented in this book that the answer will be a positive one in the end and that my intellectual journey exploring knowledge-based approaches to marketing was worthwhile.

Charles Savage will follow up on the 'So what?' question in his Afterword to this book. But before that, we will have to answer the question of what knowledge-based marketing really is about.

Acknowledgements

This book would never have been possible without the help, advice, and support of many people. I am most greatly indebted to Jiro Nonaka for his guidance and kind support of my work, not least in the Mazda and Maekawa case studies. I learned a lot from working with him and he gave invaluable feedback on my research and the manuscript of this book. I am also grateful to Nigel Holden and Charles Savage for many insightful discussions and their many helpful comments. All mistakes are solely mine.

I would also like to thank the following people: Atsushi Degawa for his invaluable feedback, especially in the early stages of the research project in 2004 and 2005; Jeff Funk who made the research stay at Hitotsubashi University in Tokyo possible in the first place; Marcus Heuberger and my former colleagues at Schindler Japan for their kind support and adivce with the Schindler case study; Kaz Ichijo for his kind support of my work, especially concerning the Toyota case study and Japanese transcriptions; Yoshio Iwasaki for his support and advice with the Maekawa case study; Helmut Kasper for his feedback and support; Jürgen Mühlbacher for his feedback and advice; Kazuo Mukai for the many interesting discussions and his kind support with the HP case study; Hiroyuki Ogamoto for his helpful suggestions and his kind support with the Siemens case study; and Alfred Taudes for his feedback, advice, and support.

Others, most importantly Michael O. B. Krähe, also helped and supported me on my research endeavour in Japan, introducing interesting companies/interviewees, correcting my Japanese, and helping with the Japanese and English transcriptions, and so on. I would very much like to thank everybody, even though I cannot list all of their names here.

My parents Annemarie and Gerhard and my girlfriend Claudia have been a splendid reservoir of vital energy and I am most grateful for their patience, love, and encouragement.

Last but not least, special thanks go to the following people at Palgrave Macmillan, for their support and excellent work in the publication process: Stephen Rutt, Virginia Thorp, Mirabelle Boateng, and Jo North.

Abbreviations

CKM	customer knowledge management
CoP	community of practice
CRM	customer relationship management
DCM	demand chain management
GM	General Motors
HRM	human resource management
IMV	innovative international multi-purpose vehicle project
IJV	international joint venture
JV	joint venture
KAM	key account management
KC	knowledge community
LC	learning community
MNC	multinational corporation/company
PDM	product development management
R&D	research and development
SCM	supply chain management
TPCA	Toyota Peugeot Citroën Automobile Czech
TQM	total quality management
USP	unique selling proposition

1
Introduction

'In an economy where the only certainty is uncertainty, the one sure source of lasting competitive advantage is knowledge' (Nonaka, 1991: 96). This statement was the starting point of Ikujiro Nonaka's seminal article on the knowledge-creating company (cf. also 3.4) in 1991. According to Wenger and Snyder (2000: 139), '[t]oday's economy runs on knowledge, and most companies work assiduously to capitalize on that fact', and Davis and Botkin (1994: 165) posit that '[t]he next wave of economic growth is going to come from knowledge-based businesses'. Indeed, significant numbers of scholars have observed that our society has evolved into a 'knowledge society'[2] (cf., for example, Bell, 1973; Drucker, 1969, 1993; Stehr, 1994; Stehr and Ericson, 1992; Toffler, 1990), and our economy into a 'knowledge economy' (cf., for example, Bertels and Savage, 1999; Burton-Jones, 1999; Drucker, 2002; Leibold, Probst, and Gibbert, 2002; Mokyr, 2002; Skyrme, 1999; Teece, 2000a).

In this knowledge society, 'knowledge is *the* primary resource for individuals and for the economy overall. Land, labor, and capital – the economist's traditional factors of production – do not disappear, but they become secondary. They can be obtained, and obtained easily, provided there is specialized knowledge' (Drucker, 1992: 95, original emphasis). But this knowledge society is also a society of organizations, with 'the purpose and function [...] [being] the integration of specialized knowledge into a common task' (Drucker, 1992: 96). In fact, Davenport (2005: 9) concludes that 'the most important processes for organizations today involve knowledge work', and the core message of an earlier work was that 'the only sustainable advantage a firm has comes from what it collectively knows, how efficiently it uses what it knows, and how readily it acquires and uses new knowledge'

(Davenport and Prusak, 2000: xxiv). Put in a nutshell, knowledge is a company's only enduring source of advantage in an increasingly competitive world (Birkinshaw, 2001). Dealing with knowledge creation, transfer, and exploitation will be more and more critical to the survival and success of corporations, and of societies (Hedlund and Nonaka, 1993). This ever increasing importance of knowledge for organizations has raised – and still raises – 'questions about how organizations process knowledge and, more importantly, how they create new knowledge' (Nonaka, 1994: 14). Indeed, as Ichijo and Nonaka (2006: 3) note, 'the success of a company in the twenty-first century will be determined by the extent to which its leaders can develop intellectual capital through knowledge creation and knowledge-sharing on a global basis' as knowledge constitutes a competitive advantage in this age.

What followed was a knowledge and knowledge management boom. As a matter of fact, knowledge management has not only become a ubiquitous phenomenon both in the academic and in the corporate world, but it has also turned into one of the most prominent and widely discussed management concepts of the postmodern era. Publications on knowledge management are legion, and business practitioners do not fail to stress its importance for the competitiveness of their corporations. Prusak (2001: 1002) – who sees 1993 as the beginning of knowledge management – argues that knowledge management is 'a practitioner-based, substantive response to real social and economic trends', with the three most important ones being globalization, ubiquitous computing, and the knowledge-centric view of the firm. Even though knowledge management has also been analysed and discussed as a management fad and within the framework of management fashion models (cf., for example, Scarbrough, Robertson, and Swan, 2005; Scarbrough and Swan, 2001; Skyrme, 1998) to explain its diffusion and 'strong rhetorical appeal' (Alvesson, Kärreman, and Swan, 2002: 282), no management scholar or practitioner is likely to disagree with Newell and fellow researchers' (2002: 2) pronouncement to the effect that '[m]anaging knowledge and knowledge workers is arguably the single most important challenge being faced by many kinds of organizations across both the private and public sectors in the years to come'. Indeed, it is now widely recognized that the effective management of knowledge assets is a key requirement for securing competitive advantage in the knowledge economy (Boisot, 1998). Building mainly from the theory of organizational knowledge creation (3.4), and communities of practice (3.5), I will look at how organizations process and create knowledge in Chapter 3.

Ikujiro Nonaka's publications (for example, Nonaka, 1991, 1994, 2005; Nonaka and Takeuchi, 1995) have drawn attention to Japanese firms as – particularly effective – knowledge-creating companies, a feature that supposedly helped them to create the dynamics of innovation and to become world leaders in the automotive and electronics industries, among others, in the 1980s and the beginning of the 1990s. The difference, it was argued, between Japanese and Western firms, lies in the focus on tacit knowledge of the former and that on explicit knowledge of the latter (Hedlund and Nonaka, 1993; Nonaka and Takeuchi, 1995; Nonaka, von Krogh, and Voelpel, 2006; Takeuchi and Nonaka, 2000); Japanese firms' particular ability for knowledge creation and harnessing tacit knowledge has also been recognized and discussed by Western scholars (for example, Baumard, 1999; Cohen, 1998; Davenport and Prusak, 2000; Holden, 2002; Leonard, 1998).

In recent years, knowledge management has also become a dominant area in strategic management and has increasingly been adapted to the global context. Especially the capability of multinational corporations (MNCs) to create and efficiently transfer and combine knowledge from different locations around the world is becoming more and more important as a determinant of competitive advantage and has become critical to their success and survival (Asakawa and Lehrer, 2003; Bartlett and Ghoshal, 2002; Chini, 2004; Desouza and Awazu, 2005b; Doz, Santos, and Williamson, 2001; Gupta and Govindarajan, 2000a; Macharzina, Oesterle, and Brodel, 2001; Schulz and Jobe, 2001). But despite the strong interest in and the large number of publications on the issue of knowledge flows within MNCs, the literature is 'still in the early stages of understanding the central aspects, mechanisms, and contextual factors in the process of managing knowledge in MNCs' (Foss and Pedersen, 2004: 342). In fact, 'rather little is known about the determinants of intra-MNC knowledge flows in spite of their obvious importance to theoretical arguments about the MNC' (Foss and Pedersen, 2002: 52). So far, the extant literature has mainly focused on the issue of transferring knowledge between different units in MNCs – that is, the knowledge flows within MNCs – and factors influencing it (for example, Foss and Pedersen, 2002; Gupta and Govindarajan, 1991, 2000a; Martin and Salomon, 2003; Minbaeva, Pedersen, Björkman, Fey, and Park, 2003; Mudambi, 2002). But research on the process of knowledge creation within MNCs is still scarce. This cross-border synergistic process of joint knowledge creation – 'global knowledge creation' (Nonaka, 1990b) – will be dealt with in Chapter 3.6.

Even though 'marketing functions lend themselves particularly well to an investigation of knowledge transfer within MNCs', 'there is a dearth of research on knowledge transfer in the field of marketing' and 'it is high time to include marketing in the research agenda for knowledge management and to overcome the paradox that marketing functions are neglected in the discussion on knowledge transfer' (Schlegelmilch and Chini, 2003: 220–1, 226). Despite the obvious importance of knowledge to the marketing discipline, the marketing literature has struggled for more than ten years to come to an understanding of the nature of marketing knowledge and there does not seem to be a common ground unifying scholars (Kohlbacher, Holden, Glisby, and Numic, 2007). Indeed, even though 'marketers have been using knowledge management long before this latter phrase began to be popularised in the management literature' (Chaston, 2004: 22–3) there has to date been 'no clear statement about the forms that marketing knowledge can take, or its content' (Rossiter, 2001: 9), and Chaston's (2004) book on knowledge-based marketing is one of the few works – if not the only one – that shows how knowledge can be utilized to underpin and enhance the marketing management function within organizations. The status quo of marketing knowledge (management) research will be reviewed in Chapter 4.

This book has been inspired and influenced by the above-mentioned major themes of knowledge as an important source of competitive advantage, of managing knowledge, global aspects of managing knowledge, and – most importantly – the role of knowledge and knowledge management in marketing management. Building from a comprehensive empirical study and the state-of-the-art literature in the field, I will introduce and define the concept of knowledge-based marketing and propose a shift towards a new dominant logic – namely a knowledge-based one – for marketing.

2
Aims of the Book and Research Questions

Based on the issues touched on in the Introduction, this chapter briefly describes the objective and aims of this book, posits the research question underlying the research project, shows what practical and theoretical implications can be expected, and finally outlines the structure of the book.

2.1 Objective and aims

This book is essentially about knowledge and knowledge creation. Its aim is not only to illustrate, analyse, and discuss knowledge-related processes in organizations but also to create new knowledge, that is, amend and extend existing theory and even build new theory. The issues mentioned above in the Preface and the Introduction have triggered a strong cognitive interest in knowledge-based marketing and management. Before embarking on the intellectual journey documented in this book, I briefly assisted in two research projects on knowledge management, one with a focus on knowledge transfer and organizational learning in MNCs in general and the other one with a focus on the transfer of marketing knowledge within Euro-Japanese MNCs. Many of the issues and insights that resulted from the projects have further deepened my cognitive interest and strongly influenced my own research project.

The fact that Ikujiro Nonaka's theory of organizational knowledge creation has drawn attention to the way Japanese firms create knowledge organizationally and innovate, and the strong position of the Hitotsubashi School of Knowledge Management have raised my interest in conducting research in Japan. This is the main reason why I decided to visit Hitotsubashi University for a two-year period to conduct most of my research project in Japan.

Based on the assumption that marketing is one of the most knowledge-intensive activities of a company, the focus lies on a particular type of knowledge, which has been widely neglected in past research: marketing knowledge (cf. also Chapters 1 and 4.1.2). In fact, according to Simonin (1999a), marketing knowledge and skills have yet to receive proper conceptual and empirical attention as a competency source of competitive advantage that can be transferred within multinationals. Besides, focusing on one particular type of knowledge reduces the complexity and facilitates the investigation for both researcher and the researched. Conclusions for other functions and other types of knowledge might then be drawn from the results in a further step (cf. Chapters 5 and 6).

The detailed aims of this book are as follows:

- analysing the role of knowledge in marketing, and the way it is created and managed;
- developing a useful definition of marketing knowledge;
- developing a conceptual framework and model of knowledge-based marketing.

These aims together constitute the overall objective of this book: to develop and build a theory of knowledge-based marketing and thus extend and contribute to the knowledge-based theory of the firm. In fact, taking a knowledge-based perspective of marketing, I will set out to explore a 'new frontier of knowledge management' (Desouza, 2005).

2.2   Research question

According to the qualitative research paradigm (research methodology will be discussed in the Appendix), it is only in the course of doing field research that one can find out which (research) questions can reasonably be asked and it is only at the end that you will know which questions can be answered by a study. Therefore, the preliminary research question I posited at the outset has been revised and complemented in the course of conducting the research project.

In accordance with the aims of the book, the main question of this research project is:

What is knowledge-based marketing and which types and patterns of marketing knowledge co-creation within MNCs can be identified?

This overall research question includes the following essential sub-questions (cf. also Chapter 4):

- What is marketing knowledge?
- What is its role in marketing and how is it created and managed?

2.3 Theoretical and practical implications

This book deals with knowledge-based approaches to marketing both in an intra-firm and also in an inter-firm context. The results from the empirical study as well as the theoretical argument presented here will have both theoretical and practical implications, which might be of interest to academics as well as practitioners. The results and implications will be discussed in detail in Chapters 5 and 6. In the academic field, the findings might be of potential interest to the following groups of researchers:

- researchers in the field of knowledge management;
- researchers in the field of (international) marketing and management;
- researchers in the field of organizational studies;
- researchers in the field of international business.

As far as practitioners are concerned, the following groups of managers might be the most interested:

- knowledge managers and other practitioners with related tasks;
- marketing managers in general as well as those engaged in international and cross-cultural marketing specifically;
- managers engaged in international business.

2.4 Structure and organization

Chapter 3 reviews and summarizes the theoretical background and thus builds the theoretical framework for my argument and analysis. The overall framework is the knowledge-based view of the firm (Chapter 3.3) and the main pillar for constructing a theory of knowledge-based marketing will be the theory of organizational knowledge creation (Chapter 3.4), and will be supported and complemented by the relevant literature on communities of practice (Chapter 3.5), global knowledge-based management (Chapter 3.6), and inter-organizational knowledge-based management (Chapter 3.7).

Chapter 4 reviews and summarizes the literature on marketing knowledge (4.1.2) and the state-of-the-field in knowledge-based management and organizational learning in marketing (functions) and their building blocks (4.1.1 and 4.1.3). Section 4.1.4 summarizes the findings and highlights the research gap and problematic issue of the extant literature. Chapter 4.2 presents the essence of this book, namely a conceptual framework and model of marketing knowledge (4.2.1) and knowledge-based marketing (4.2.2).

Chapter 5 presents the findings from the empirical study and depicts six explanatory case studies, which are subsequently analysed in Chapter 6.1.

Finally, Chapter 6 provides a summary and conclusion of the analysis and discussion of knowledge-based marketing and the case studies (Chapters 6.1 and 6.2). Chapter 6.3 posits the evolution towards a new dominant logic – a knowledge-based logic – for marketing. Chapter 6.4 presents the main conclusions from the research project, 6.5 briefly discusses managerial implications, and Chapter 6.6 deals with the limitations of the study and implications for future research.

The Appendix presents and discusses the research methodology underlying this research project. It deals with exploratory/qualitative research issues (A.1) and explains data collected and methods used for analysis for the empirical research project (A.2).

3
Theoretical Framework: the Knowledge-based View of the Firm

> A knowledge-based theory of the firm differs from all previous theories in that it must grasp the un-understood. (Spender and Grant, 1996: 8)

I have already referred to the prominent role of knowledge for individuals, organizations, and society as a whole. Building on the theoretical framework of the knowledge-based view of the firm, this prominent role of knowledge, along with key issues related to the creation, sharing, transfer, and storage of knowledge, will be further explored in this chapter.

3.1 Knowledge

Before starting to discuss the management of knowledge and related issues, it is necessary to define what is meant by knowledge. As a matter of fact, the discussion of knowledge is not a new one and basically derives from a long philosophical tradition but the discussion also draws from many other fields such as sociology, psychology, and economics (for a well-referenced discussion of knowledge in both the Western and Japanese traditions, see Nonaka and Takeuchi, 1995).

In the relevant knowledge management literature (in business administration), a distinction between data, information, and knowledge – or at least between information and knowledge – has regularly been made (for example, Baumard, 1999; Davenport and Prusak, 2000; Dixon, 2000; Nonaka and Takeuchi, 1995; Tsoukas and Vladimirou, 2001). Data can be defined as 'a set of discrete, objective facts about events' and in an organizational context data are most usefully explained as 'structured records of transactions' (Davenport and Prusak, 2000: 2).

Information has frequently been described as a message or a flow of messages and it can be thought of as data that make a difference (for example, Davenport and Prusak, 2000; Nonaka and Takeuchi, 1995). Knowledge refers to information embedded in the context of system-specific patterns of experience and is always for a specific purpose. Wiig (2004: 337) contends that knowledge is used to 'interpret information about a particular circumstance or case to handle the situation' and that knowledge is about 'what the facts and information mean in the context of the situation'. According to Nonaka and Takeuchi (1995: 58), knowledge is created by the flow of information, anchored in the beliefs and commitment of its holder and is therefore essentially related to human action. Dixon (2000) uses the term 'common knowledge' to differentiate the knowledge that employees learn from doing the organization's tasks from book knowledge or from lists of regulations or databases of customer information. In this sense, 'common knowledge is the "know how" rather than the "know what" of school learning' (Dixon, 2000: 11). Holden (2002: 65) seems to agree on that when he emphasizes that 'in the management context "knowledge" means organizational knowledge rather than the contents of encyclopædias or reference books'. According to Dixon (2000: 13) knowledge is defined 'as the meaningful links people make in their minds between information and its application in action in a specific setting'. Davenport and Prusak (2000: 5) offer a very useful definition of knowledge, making clear that knowledge is not neat or simple:

> Knowledge is a fluid mix of framed experience, values, contextual information, and expert insight that provides a framework for evaluating and incorporating new experiences and information. It originates and is applied in the minds of knowers. In organizations, it often becomes embedded not only in documents or repositories but also in organizational routines, processes, practices, and norms.

Knowledge can furthermore be divided into declarative and procedural knowledge. Declarative knowledge is about describing something, that is, declarative knowledge is knowledge about facts and concepts, as it deals with information about a situation (Haghirian, 2003; Zack, 1999a). Procedural knowledge, on the contrary, refers to 'know-how' to perform a certain task or activity, i.e. a procedure that represents embedded experience and successful solutions to complex tasks, as well as a co-ordination of solutions among various tasks in the organization (Haghirian, 2003). Accordingly procedural knowledge deals with information about how something occurs or is performed and it is based on distinct systems and derives from past planning of action sequences that were successful (Haghirian, 2003; Zack, 1999a).

Tsoukas and Vladimirou (2001: 979, original emphasis) come to the following conclusion on what knowledge is: '*knowledge is the individual ability to draw distinctions within a collective domain of action, based on an appreciation of context or theory or both*'. According to them, 'such a definition of knowledge preserves a significant role for human agency, since individuals are seen as being inherently capable of making (and refining) distinctions, while also taking into account collective understandings and standards of appropriateness, on which individuals necessarily draw in the process of making distinctions, in their work' (Tsoukas and Vladimirou, 2001: 979). Last but not least, Wiig (2004: 336) offers the following operational definition of knowledge: 'The content of understanding and action patterns that govern sense-making, decision making, execution, and monitoring'. According to him, knowledge 'consists of facts, perspectives and concepts, mental reference models, truths and beliefs, judgments and expectations, methodologies, and know-how' (Wiig, 2004: 337).

In this book – since it heavily draws and builds upon Nonaka's theory of organizational knowledge creation, see 3.4 – I shall follow Nonaka and Takeuchi's (1995: 58, original emphasis) definition of knowledge as '*a dynamic human process of justifying personal belief toward the "truth"*'. In this context, it is crucial to differentiate between explicit and tacit knowledge, and Nonaka and Takeuchi (1995) draw on Polanyi's (1966) distinction between tacit knowledge and explicit knowledge. They view tacit knowledge as 'personal, context-specific, and therefore hard to formalize and communicate' and explicit – or codified – knowledge as knowledge that 'is transmittable in formal, systematic language' (Nonaka and Takeuchi, 1995: 59). Indeed, explicit knowledge is formal and systematic and can be easily communicated and shared with others, while tacit knowledge refers to a kind of knowledge which is highly personal, hard to formalize, and thus difficult to communicate to others, as it is deeply rooted in action (Nonaka, 1996: 21).

3.2 Knowledge management and organizational learning

As mentioned in the Introduction, in contemporary organizations significant emphasis is placed on the processes of knowledge creation, sharing, and learning (for example, Barrett, Cappleman, Shoib, and

Walsham, 2004; Nonaka and Takeuchi, 1995; Senge, 1990). Organizational learning, the learning organization, and knowledge management have emerged as seminal concepts for both academe and practitioners and have received ample attention (for example, Buckman, 2004; Davenport and Prusak, 2000; Dixon, 2000; English and Baker, 2006; Fiol and Lyles, 1985; Garvin, 1993, 2003; Huber, 1991; Leonard, 1998; von Krogh, Ichijo, and Nonaka, 2000). As Easterby-Smith and Lyles (2003: 1) note in their handbook of organizational learning and knowledge management, '[t]he fields of organizational learning and knowledge management have developed quickly over the last decade, and the academic literature has demonstrated increasing diversity and specialization'. Finally, integrative frameworks for both concepts have been put forward, thus helping to reduce conceptual confusion and facilitate communication between researchers who have treated them separately for a long time (for example, Vera and Crossan, 2003). Since knowledge management 'has started to emerge as an area of interest in academia and organisational practice' in the middle of the 1990s (McAdam and McCreedy, 1999: 91), knowledge has frequently been identified as a crucial strategic resource and asset for corporations (Earl, 1997; Lyles and Schwenk, 1992; Probst, Büchel, and Raub, 1998), and strategies for knowledge creation and management have been proposed (Choo and Bontis, 2002b; Hansen, Nohria, and Tierney, 1999; Hofer-Alfeis and van der Spek, 2002; Teece, 2000b; Un and Cuervo-Cazurra, 2004; Zack, 1999b). Indeed, in the knowledge economy with its diminishing returns, 'knowledge management can be an important component of competitive strategy, as it will assist the firm in pushing the limits of its business model' (Teece, 2000b: 49).

This book focuses mainly on the knowledge management literature and sees organizational learning as one of several processes of knowledge-based management. Basically, I adopt the classic definition of organizational learning by Fiol and Lyles (1985: 803) as 'the process of improving actions through better knowledge and understanding' because it puts knowledge at the heart of the learning process. A learning organization is then 'an organization skilled at creating, acquiring, interpreting, transferring, and retaining knowledge, and at purposefully modifying its behavior to reflect new knowledge and insights' (Garvin, 2003: 11, removed emphasis).

As I mentioned in the Preface, the term 'knowledge management' needs to be used carefully to avoid misunderstandings, especially if we do *not* want the term to mean information and document management and IT-based tools for collecting, storing, and searching documents. An extensive review of the relevant literature as well as the empirical study confirmed this concern, as indeed the 'vast majority of texts on knowledge management tend to focus on the information technology (IT) aspects of managing the concept' (Chaston, 2004: x) and practitioners seem to have absorbed this view as well. However, as Chaston (2004: x) correctly notes, '[a]lthough management of technology is critical, there is an equally important need for the provision of materials describing how knowledge can be utilized in the execution of functional management tasks'. In fact, '[t]hough we have seen a tendency – especially among vendors of software – to reductively define knowledge management as moving data and documents around, knowledge management grew out of an understanding of the critical value of these other, less digitized factors, and the clear need to devise ways to support and benefit from them' (Prusak, 2001: 1003). Therefore, knowledge management should not be limited to a function of merely collecting and documenting, but should rather actively utilize information as a resource, and process, prepare, and format it in such a way that it becomes relevant organizational knowledge. In order to distinguish between the narrow meaning of knowledge management – as data and information management – and the broader and more comprehensive way of dealing with knowledge and knowledge assets in firms, I will use the term 'knowledge-based management' for the latter approach. This is also consistent with the terminology of the knowledge-based theory of the firm, which will be discussed in the next section.

3.3 The knowledge-based theory of the firm

According to Teece (2000b: 42), the 'modern corporation, as it accepts the challenges of the new knowledge economy, will need to evolve into a knowledge-generating, knowledge-integrating and knowledge-protecting organization'. As mentioned in the Introduction, Prusak (2001: 1002) argues that knowledge management is 'a practitioner-based, substantive response to real social and economic trends', one of which was the knowledge-centric view of the firm. In this view, a firm is 'seen as a coordinated collection of capabilities', and the 'main building block of these capabilities ... is knowledge, especially the knowledge that is mostly tacit and specific to the firm' (Prusak, 2001: 1003). The knowledge-based view or knowledge-based theory of the firm was comprehensively discussed in a special issue of the *Strategic Management Journal* edited by Spender and Grant in 1996 (Grant, 1996b;

Spender, 1996b; Spender and Grant, 1996). In this special issue, the editors – in selecting the papers – 'have sought to move toward the still hidden knowledge-based theory of the firm' (Spender and Grant, 1996: 9).

By now, the knowledge-based theory of the firm 'has arguably established itself as the mainstream literature informing the discourse on knowledge in organizations' (Patriotta, 2003: 25). It has been influenced by the work of Penrose (1995) and – more generally – by the so-called resource-based view of the firm (for example, Acedo, Barroso, and Galan, 2006; Barney, 1991, 2001; Peteraf, 1993; Wernerfelt, 1984, 1995). Indeed, Grant (1996b: 110) sees the knowledge-based view as 'an outgrowth of the resource-based view', a view that is also echoed by other scholars in the field (for example, Patriotta, 2003).

The knowledge-based theory of the firm criticizes the resource-based view of the firm and tries to overcome the weaknesses of this approach: the resource-based view of the firm looks inside firms in terms of the resources they own (Nonaka, Toyama, and Nagata, 2000), and according to this view, a firm is a collection of resources, and those with superior resources will earn rents (for example, Barney, 1991, 2001; Conner, 1991; Foss, 1997; Itami and Roehl, 1987; Mahoney and Pandian, 1992; Peteraf, 1993; Wernerfelt, 1984, 1995). The resource-based view treats knowledge as one such resource, but empirical and theoretical research on the resource-based view of the firm so far has been mainly focused on how firms keep their unique resources and resulting competitive advantages through such conditions as imperfect substitutability and limited mobility of resources (Nonaka, Toyama, and Nagata, 2000: 7–8; Nonaka and Toyama, 2003: 4; cf. also Amit and Schoemaker, 1993; Barney, 1991; Nonaka, von Krogh, and Voelpel, 2006; Peteraf, 1993; Wernerfelt, 1984). Therefore, Nonaka and Toyama (2003: 4) conclude that – although it deals with the dynamic capability of the firm – 'the resource-based view of the firm fails to explain the dynamism in which the firm continuously builds such resources through interactions with the environment' (cf. also Nonaka, Toyama, and Nagata, 2000: 7). 'What is missing in the resource-based approach is a comprehensive framework that shows how various parts within the organization interact with each other over time to create something new and unique' (Nonaka and Takeuchi, 1995: 49). As a result, the 'knowledge-based theory of the firm can yield insights beyond the production-function and resource-based theories of the firm' and is 'a platform for a new view of the firm as a dynamic, evolving, quasi-autonomous system of knowledge production and application' (Spender, 1996b: 59).

The knowledge-based theory of the firm also draws upon other research streams including epistemology, organizational learning, organizational capabilities, innovation, and new product development (Burton-Jones, 1999). According to Patriotta (2003), the idiosyncratic knowledge base underlying a firm's performance includes resources (Barney, 1991; Penrose, 1995; Peteraf, 1993; Wernerfelt, 1984), routines (Nelson and Winter, 1982), competencies (Prahalad and Hamel, 1990), capabilities (Amit and Schoemaker, 1993; Eisenhardt and Martin, 2000; Kogut and Zander, 1992; Leonard-Barton, 1992; Teece, Pisano, and Shuen, 1997), and intellectual capital (Nahapiet and Ghoshal, 1998; Quinn, 1992). Obviously, various different research streams can be subsumed under the heading of the knowledge-based view of the firm, but even though each of these streams has a distinct focus, they basically all share the notion that knowledge is the critical source of competitive advantage for firms. A study by Acedo, Barroso, and Galan (2006) confirmed the links among the resource-based view, the knowledge-based view, and the dynamic capability perspective.[3] However, they see the knowledge-based view as one main trend within the resource-based theory of the firm and identify two large subgroups of the knowledge-based view: one – which is closer to the resource-based view group – asserts that knowledge is the most important strategic resource for organizations (Conner and Prahalad, 1996; Grant, 1996b; Kogut and Zander, 1992) and the other subgroup maintains a less positivist view of knowledge analysis and adopts a more pluralistic epistemology, redolent of social constructivism (Spender, 1996b; Tsoukas, 1996). Interestingly, each branch of the knowledge-based view is defended by the editors of the special issue mentioned above, Grant and Spender (Acedo, Barroso, and Galan, 2006). Grant (1996b) acknowledges that the two different approaches originate from their different academic backgrounds: economics (Grant) and philosophy, psychology, and technology (Spender). Acedo Barroso and Galan (2006) further argue that the work of Teece, Pisano, and Shuen (1997) lies (somewhere) in between these two views. A detailed discussion and comparison of the different research streams would go beyond the scope of this book. I therefore adopt here the notion of the knowledge-based view of the firm in accordance with the work by Nonaka and associates (Nonaka and Toyama, 2002, 2005; Nonaka, Toyama, and Nagata, 2000).

According to Nonaka, Toyama, and Nagata (2000b: 1), the knowledge-based view of the firm is the most recent development in the theory of the firm and 'views a firm as a knowledge-creating entity, and argues that knowledge and the capability to create and utilize such knowledge are the most important source of a firm's sustainable competitive advantage' (cf. also Cyert, Kumar, and Williams, 1993; Kogut and Zander, 1996; Metcalfe and James, 2000; Nahapiet and Ghoshal, 1998; Nelson, 1991; Nonaka, 1991, 1994; Nonaka and Takeuchi, 1995; Prahalad and Hamel, 1990). Indeed, 'in the view of the firm as a knowledge-creating entity, a firm is a dynamic entity which actively interacts with its environment, and reshapes the environment, and even itself, through the process of knowledge creation' (Nonaka and Toyama, 2005: 420). According to Grant (1997: 454), the 'knowledge-based view promises to have one of the most profound changes in management thinking since the scientific management revolution' of the early decades of the twentieth century. The knowledge-based theory of the firm has also been 'strongly influenced by growing recognition of different types of knowledge and their characteristics' (Spender and Grant, 1996: 8). Indeed, the work of Polanyi (1962, 1966) and Nelson and Winter (1982), has been especially influential in directing attention to knowledge which is embodied in individual and organizational practices and cannot be readily articulated. But such knowledge is of critical strategic importance because, unlike explicit knowledge, it is both inimitable and appropriable (Spender and Grant, 1996).

Knowledge-based approaches in management research portray organizations as primary vehicles for producing, transferring, and combining knowledge (cf., for example, Grant, 2002; Kogut and Zander, 1996; Nonaka, 1994) and basically see firms as social communities that serve as efficient mechanisms for the creation and transformation of knowledge into economically rewarded products and services (Kogut and Zander, 1993).[4] According to Grant (1996b: 112), the assumption that the critical input in production and primary source of value is knowledge is fundamental to a knowledge-based theory of the firm. Spender (1989: 33) redefined the organization 'as the set of ideas which influence individual behavior' and sees the firm as 'a body of knowledge, what might now be called a "knowledge-base"'. Therefore, 'it is the firm's knowledge, and its ability to generate knowledge, that lies at the core of a more epistemologically sound theory of the firm' (Spender, 1996b: 46). Spender (1996b: 47) calls for such a 'knowledge-based theory in which organizations are enduring alliances between independent knowledge-creating entities, be they individuals, teams or other organizations, and tangible resources are subordinated to the services they provide'. Firms have also been identified as 'knowledge systems' (Grant, 1996a) or 'distributed knowledge systems' (Tsoukas, 1996) as well as a 'repository for knowledge' (Teece, 1998) by scholars in the field. In a similar vein, Patriotta (2003: 25) concludes that the most distinctive trait of the knowledge-based theory of the firm is 'the conceptualization of the firm itself as a body of knowledge'.

A central question for the theory of the firm is, 'Why do firms differ?' (Nelson, 1991; Nonaka and Toyama, 2005), and in trying to understand business and economics we also keep coming back to the questions of 'What is a firm?' and 'How does it function?' (Nonaka and Toyama, 2002). This is in contrast to Tsoukas (1996) who argues that from a research point of view, what needs to be explained is not so much 'why firms differ' – according to him they inevitably do – but rather what are the processes that make them similar. The answer provided by Nonaka and Toyama (2005: 420) is that 'firms differ because they want and strive to differ' and that 'in order to explain why firms differ we have to deal with the subjective elements of management, such as management vision, the firm's value system, and the commitment of employees'. In fact, in organizational knowledge creation, it is such differences in human subjectivities that help create new knowledge (Nonaka and Toyama, 2005: 421). The knowledge-based view of the firm is therefore different from the positioning school and the resource-based view of the firm (see above and also Patriotta, 2003). The positioning school mainly focuses on the environment in which the organization operates (Porter, 1980) and explains firm differences with reference to the difficulties in entering an industry or a strategic group (Nonaka and Toyama, 2003, 2005).

The knowledge-based theory of the firm has also helped to raise important questions about sustainability of competitive advantages and cumulative strategic change within the organization (Choi and Lee, 1997) and deals with the importance of knowledge within the corporation. According to Nonaka, Toyama, and Nagata, (2000: 2), '[k]nowledge and skills give a firm a competitive advantage because it is through this set of knowledge and skills that a firm is able to innovate new products/processes/services, or improve the existing ones more efficiently and/or effectively'. However, it is increasingly difficult for firms to attain and sustain competitive advantages through the reallocation of capital (Bresman, Birkinshaw, and Nobel, 1999). As Hansen and Nohria (2004: 22) put it, the ways for MNCs to compete successfully by exploiting scale and scope economies or by taking advantage of imperfections in the world's goods, labour, and capital markets are no longer as profitable as they once were, and as a result, 'the new economies of scope are based on the ability of business units, subsidiaries and functional departments within the company to collaborate successfully by sharing knowledge and jointly developing new products and services'. In addition, contexts of competition and international business are changing also, which causes uncertainty for organizations and puts pressure upon them to change and renew their existing practices (Choi and Lee, 1997). This is also why Spender and Grant (1996: 9) conclude their introduction to the special issue by summarizing Spender (1996b) with the following:

> the knowledge-based theory of the firm is a paradigmatic gateway, the point in the evolution of our field where we abandon the older concept of a theory as a blue-print for creating the firm, and move towards a more agricultural notion of management as the intervention in and husbandry of the natural knowledge-creating processes of both individuals and collectivities, be they societies as they create and are reconstituted by their culture, or firms as they create and are reconstituted by their creations.

The knowledge-based view of the firm suggests that 'knowledge creation and management are key in today's knowledge intensive society' (Hanvanich, Dröge, and Calantone, 2003: 124). In line with this, Nonaka and fellow researchers interpret the knowledge-based theory of the firm as a 'knowledge-creating view of the firm' with the raison d'être of a firm being to continuously create knowledge (Nonaka and Toyama, 2002, 2005; Nonaka, Toyama, and Nagata, 2000). They argue that the knowledge-creating view of the firm 'is different from other theories of the firm in its basic assumptions that humans and organizations are dynamic beings, and in its focus on the process inside the firm' (Nonaka, Toyama, and Nagata, 2000: 2). Zack (2003: 69) found four characteristics of a knowledge-based organization, namely process, place, purpose, and perspective and argues that the knowledge-based organization 'is a collection of people and supporting resources that create and apply knowledge via continued interaction'. In a similar vein, business organizations can be seen as collections of knowledge assets, and the integration, updating, maintenance, and management of those assets are of great importance (Tsoukas and Mylonopoulos, 2004). According to Teece (1998: 75), the 'essence of the firm is its ability to create, transfer, assemble, integrate, and exploit knowledge assets' and knowledge assets (cf. 3.4.3) 'underpin competences, and competences in turn underpin the firm's product and service offerings to the market'. In fact, 'competitive advantage (superior profitability) at the enterprise level depends upon the creation and exploitation of difficult-to-replicate non-tradable assets, of which knowledge assets are the most important' (Teece, 2000b: 44).

3.4 The theory of organizational knowledge creation

The theory of organizational knowledge creation has basically been developed by Ikujiro Nonaka and fellow researchers (cf., for example, Nonaka, 1991, 1994, 2005; Nonaka and Konno, 1998, 2003; Nonaka and Takeuchi, 1995; Nonaka and Toyama, 2002, 2003, 2005; Nonaka, Toyama, and Konno, 2000; Nonaka, Toyama, and Nagata, 2000; Nonaka, von Krogh, and Voelpel, 2006). The 1995 book (Nonaka and Takeuchi, 1995) is regarded as 'one of the most cited theories in the knowledge management literature' (Choo and Bontis, 2002b: 11; cf. also Choo, 2003; Takeuchi and Nonaka, 2004). Discussing the whole theory in detail – Nonaka has been developing this theory for more than twenty years now – would go beyond the scope of this book. In the following sections, its main aspects and concepts will be presented briefly and concisely.

3.4.1 The knowledge-creating company

Nonaka's publications have drawn attention to Japanese firms as knowledge-creating companies, a feature that supposedly helped them to create the dynamics of innovation and to become world leaders in the automotive and electronics industries, among others, in the 1980s and the beginning of the 1990s. Generally speaking, the difference between Japanese and Western firms lies in the former's strength of leveraging tacit knowledge while the latter tend to focus rather on explicit knowledge (Hedlund and Nonaka, 1993; Nonaka and Takeuchi, 1995; Nonaka, von Krogh, and Voelpel, 2006; Takeuchi and Nonaka, 2000).[5] Japanese firms' particular aptitude for knowledge creation and harnessing tacit knowledge has also been recognized and credited by Western scholars (for example, Baumard, 1999; Cohen, 1998; Davenport and Prusak, 2000; Holden, 2002; Leonard, 1998). According to Burton-Jones (1999: 31, original emphasis), the 'main point in the long run ... is that only *tacit* knowledge, whether alone or in conjunction with explicit knowledge, can give a firm a sustainable competitive advantage'.

In fact, this distinction between tacit and other types of knowledge is widely accepted among knowledge management researchers (for example, Spender, 2003; von Hippel, 1994). This is also closely related to two different paradigms in organizational theory and management practice: the information-processing paradigm which leads to a rather technical concept of knowledge management focusing on information technology (IT) and explicit knowledge, and the knowledge-creation paradigm which emphasizes intellectual capability and human creativity and tacit knowledge (Ichijo, 2002, 2004). According to Leonard (1998: 10), for instance, knowledge management 'demands the ability to move knowledge in all directions – up, down, across', which is why we often talk of knowledge flows within organizations, and Wiig (2004: 338) defines knowledge management as, '[t]he systematic, explicit, and deliberate building, renewal, and application of knowledge to maximize an enterprise's knowledge-related effectiveness and returns from its knowledge and intellectual capital assets'. In contrast to that, Nonaka and Takeuchi (1995: 3) mean 'the capability of a company as a whole to create new knowledge, disseminate it throughout the organization, and embody it in products, services, and systems' by organizational knowledge creation and develop a dynamic model of this process (SECI model, see below). Indeed, '[s]ince knowledge is socially constructed, focus on knowledge creation, rather than knowledge transfer, becomes paramount for organizational learning' (Plaskoff, 2003: 164). Hence, managing existing knowledge alone is simply not enough (Umemoto, 2002).

The focus on creating new knowledge rather than merely managing existing knowledge within a firm is one of the most important contributions of Nonaka's theory. Another one is the analysis of the process of organizational knowledge creation rather than solely the creation and application of knowledge by individuals. This is in contrast to Grant (1996b: 112) who works with the assumptions that knowledge creation is an individual activity and that the primary role of firms is in the application of existing knowledge to the production of goods and services. However, Grant (1996b: 121) also acknowledges that 'a more comprehensive knowledge-based theory of the firm will embrace knowledge creation and application'.

But even though this distinction between tacit and explicit knowledge is widely accepted among knowledge management researchers, it is important to note that they are not distinct categories as knowledge exists on a spectrum and all knowledge has tacit dimensions (cf., for example, Dixon, 2000; Leonard and Sensiper, 1998; Leonard and Swap, 2005a; Polanyi, 1966; Tsoukas, 1996). Tsoukas (1996: 14, original emphasis) puts it like this: 'Tacit knowledge is the necessary component of *all* knowledge; it is not made up of discrete beans which may be ground, lost or reconstituted.' Finally, the notion of knowledge as a continuum emphasizes the contrasting natures of tacit and explicit knowledge, and their interaction (Nonaka and Peltokorpi, 2006; cf. also Cavusgil, Calantone, and Zhao, 2003).

Based on the assumption that tacit and explicit knowledge are not totally separate but mutually complementary entities and that knowledge is created through the interaction between tacit and explicit knowledge, Nonaka (1994) proposed a model of four different modes of knowledge conversions (cf. also Nonaka and Takeuchi, 1995): from tacit knowledge to tacit knowledge (socialization), from tacit knowledge to explicit knowledge (externalization), from explicit knowledge to explicit knowledge (combination), and from explicit knowledge to tacit knowledge (internalization). The knowledge-creation process starts with the accumulation of personal, hard-to-externalize, subjective, and contextual tacit knowledge, which is then converted through the phases of socialization, externalization, combination, and internalization (SECI) into more objective explicit knowledge (Nonaka, 1991, 1994; Nonaka and Takeuchi, 1995). This model is widely accepted and has long become state-of-the-art in the theory of knowledge management and creation (for critical literature see 6.6). Figure 3.1 shows the detailed model of the SECI process. I will not discuss it in detail here, but I will come back to it when discussing knowledge-based marketing later on (cf. specifically 4.2.3.).

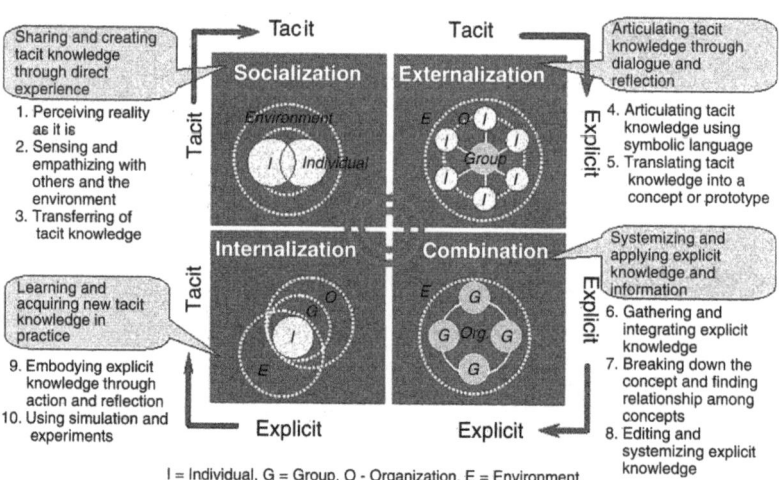

Figure 3.1 The SECI model of knowledge creation (from Nonaka and Toyama, 2003: 5)

3.4.2   Basic components of the knowledge-based firm

The theory of organizational knowledge creation has been further extended and refined by integrating the concepts of context/place (*ba*) (Nonaka and Konno, 1998; Nonaka and Toyama, 2003; Nonaka, Toyama, and Konno, 2000) and leadership and by identifying enabling conditions as well as certain barriers for knowledge creation (Ichijo, 2004; von Krogh, Ichijo, and Nonaka, 2000). In yet further extensions of the knowledge-based theory of the firm (Nonaka and Toyama, 2005), knowledge creation is described through the shared context of interaction (*ba*), visions, driving objectives, dialogues, and practices (cf. Figure 3.2), and is linked with the concept of phronetic leadership (Nonaka and Peltokorpi, 2006; Nonaka and Toyama, 2006a). In this model, the SECI process of knowledge conversion occurs through interaction between dialogues and practice, while phronetic leadership – although not indicated in the figure – influences organization-wide activities. The concept of phronetic leadership will not be dealt with further in this book.

Vision. According to Ichijo (2006b: 86) '[i]nstilling a knowledge vision emphasizes the necessity for moving the mechanics of business

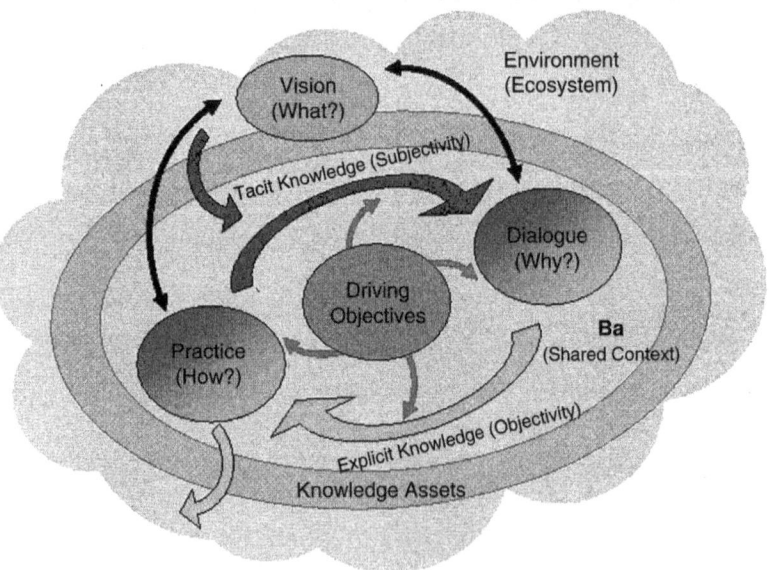

Figure 3.2 The theory of the knowledge-creating firm (from Nonaka and Toyama, 2005: 423)

strategy to creating an overall vision of knowledge in the organization'. This vision 'gives the intention via describing what knowledge should be generated' (Taudes, Trcka, and Lukanowicz, 2002: 143). The knowledge vision of a firm arises from confronting the fundamental question: 'Why do we exist?' By going beyond profits and asking 'Why do we do what we do?' the mission and domain of the firm become defined. This knowledge vision gives a direction to knowledge creation (Nonaka and Toyama, 2005). Indeed, due to the dispersed nature of organizational knowledge creation, and the need for co-ordination of teams and knowledge transfer, the theory of organizational knowledge creation emphasized the development of 'knowledge visions' in organizations (Nonaka and Takeuchi, 1995; von Krogh, Ichijo, and Nonaka, 2000). Therefore, the 'most critical element of corporate strategy is to conceptualize a vision about what kind of knowledge should be developed and to operationalize it into a management system for implementation' (Nonaka and Takeuchi, 1995: 74). Knowledge visions specify a 'potentiality for being' and both result from and inspire conversations and rhetoric throughout the organizations, and as such they represent important resources for justification involved in organizational knowledge creation (Nonaka, Peltokorpi, and Tomae, 2005; Nonaka, von Krogh, and Voelpel, 2006; cf. also Giroux and Taylor, 2002). In fact, knowledge-based visions, which both result from and inspire dialogues throughout organizations, need to be internalized by all employees (Nonaka and Peltokorpi, 2006).

Driving objectives. As companies need to generate profits to fund knowledge creation, driving objectives, actualized in concepts, numbers, and collective discipline, orchestrate the visions, dialogues, and practices into a dynamic coherence (Nonaka and Peltokorpi, 2006; Nonaka, Peltokorpi, and Tomae, 2005). According to Nonaka and Toyama (2005), driving objectives trigger knowledge creation by questioning the essence of things. Therefore, in order to initiate constant upward spiralling knowledge creation, driving objectives must be subtle, sometimes to the point of transparency, so that the new reality can emerge through reflection and social interaction.

Dialogues. Nonaka and Toyama (2005) stress the importance of dialogues as they enhance intersubjectivity by linking *ba* (see below) within and beyond the firm's boundaries. According to Ichijo (2004), the essence of organizational activities resides in communication, which is why managing communications – encouraging active communication among organizational members – is a key enabler for knowledge creation.

Practices. Practices are 'dialectics in action', processes in which people reflect the acquired knowledge and skills based on self-transcending action (Nonaka and Toyama, 2005). In organizations, driving objectives, apprenticeships, training, and mentoring arrangements are effective ways to help new employees refine and internalize new practices, and once practices are shared and systematized throughout the company they become part of the company's knowledge assets (cf. also 3.4.3). As they are mostly tacit, they are hard to imitate by other companies and thus provide a knowledge-based competitive advantage (Nonaka and Peltokorpi, 2006).

Ba. Knowledge needs context or a physical space to be created (Nonaka and Konno, 1998; Nonaka, Toyama, and Konno, 2000; Nonaka, Konno, and Toyama, 2001). Nonaka and Toyama (2003: 6) view *ba* as 'a continuously created generative mechanism that explains the potentialities and tendencies that either hinder or stimulate knowledge creative activities'. In fact, *ba* is 'an existential place where participants share their contexts and create new meanings through interactions' and '[b]y providing a shared context in motion, *ba* sets binding conditions for the participants by limiting the way in which the

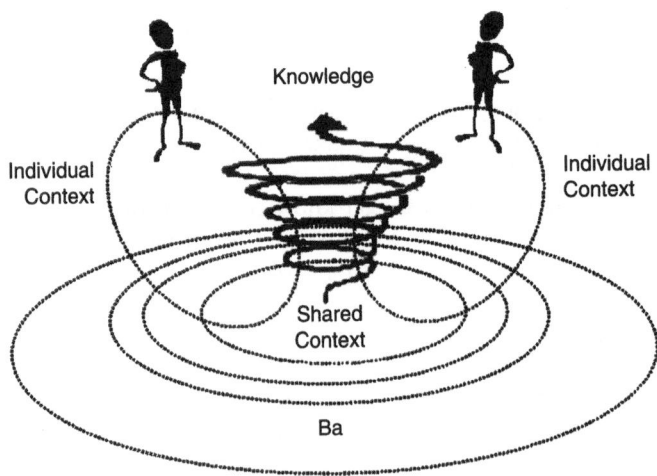

Figure 3.3 *Ba* as a shared context in motion (from Nonaka, Toyama, and Konno, 2000: 14)

participants view the world as insiders of the world' (Nonaka and Toyama, 2003: 7) (cf. also Figure 3.3). In the knowledge-based theory of the firm, a firm is considered 'a social community specializing in speed and efficiency in the creation and transfer of knowledge' (Kogut and Zander, 1996: 503; cf. also Kogut and Zander, 1992). Indeed, a firm can be conceptualized as a dynamic configuration of *ba*, i.e. as 'a collection of "ba", which interact with each other organically and dynamically' (Nonaka, Toyama, and Nagato, 2000: 9) (cf. also 4.2.2.2).

Ba is constantly evolving and thus provides a 'shared context in motion' (Nonaka and Toyama, 2003; Nonaka, Toyana, and Konno, 2000). Nonaka, Toyana, and Konno (2000: 16ff.) distinguish between four different types of *ba*: originating *ba*, dialoguing *ba*, systemizing *ba*, and exercising *ba*. But a detailed discussion would go beyond the scope of this book. Creating the right context is also one of the knowledge enablers identified by von Krogh, Ichijo, and Nonaka (2000). It examines the close connections among organizational structure, strategy and knowledge enabling and 'involves organizational structures that foster solid relationships and effective collaboration' (Ichijo, 2004: 142–3; cf. also Ichijo, 2006b).

3.4.3 Knowledge assets

According to Teece (2000b: 35), there is 'increasing recognition that the competitive advantage of firms depends on their ability to create, transfer, utilize and protect difficult-to-imitate knowledge assets' (cf. also Boisot, 1998). Nonaka, Toyama, and Nagata (2000: 14) define knowledge assets as 'inputs and outputs of knowledge-creating processes', which – unlike inputs and outputs in neoclassical economics – are often invisible, tacit, and dynamic (cf. also Kokuryo, Nonaka, and Kataoka, 2003). In fact, Figure 3.1 has already illustrated that knowledge assets are inputs and outputs of the SECI process (cf. also Nonaka and Toyama, 2005). To be precise, knowledge assets are created from the knowledge-creating process through dialogues and practices in *ba* (Nonaka and Toyama, 2005: 429). Unlike other assets, knowledge assets are intangible, are specific to the firm, and change dynamically. The essence of knowledge assets is that they must be built and used internally in order for full value to be realized, and hence cannot be readily bought and sold (Teece, 1998, 2000a, 2000b). As a result, they must be built in-house by firms, and frequently they must also be exploited internally in order for full value to be realized by the owner (Teece, 2000b: 36). Moreover, knowledge assets do not just mean the knowledge already created, such as know-how, patents, technologies, or brands, but also include the knowledge to create knowledge, such as the organizational capability to innovate (Nonaka and Toyama, 2005: 429). While knowledge assets are grounded in the experience and expertise of individuals, firms provide the physical, social, and resource allocation structure so that knowledge can be shaped into competencies (Teece, 1998: 62). Indeed, the proper structures, incentives, and management can help firms generate innovation and build knowledge assets (Teece, 2000a: 12). As a result, the competitive advantage of firms in today's economy stems not from market position, but from difficult to replicate knowledge assets and the manner in which they are deployed (Teece, 1998: 62).

Knowledge assets are then categorized into four types: experiential, conceptual, systemic, and routine knowledge assets, and they are mobilized and shared in *ba*, so that new knowledge can be continuously created (Nonaka, Toyama, and Nagata, 2000: 15–17; cf. also Nonaka, Toyama, and Konno, 2000; Kokuryo, Nonaka, and Kataoka, 2003; Umemoto, 2002).

Experiential knowledge assets. These are the shared tacit knowledge which is built through shared, hands-on experiences among organizational members and customers, suppliers, or affiliated firms. Skills and know-how, acquired and accumulated through work experiences, are examples of experiential knowledge assets. Their tacitness makes them firm-specific and difficult-to-imitate resources that provide a sustainable competitive advantage to a firm.

Conceptual knowledge assets. These are the explicit knowledge articulated through images, symbols, and language. They are the assets based on the concepts held by customers and organizational members. Since they have tangible forms, conceptual knowledge assets are easier to see than experiential knowledge assets.

Systemic knowledge assets. They are the systematized and packaged explicit knowledge, such as explicitly stated technologies, product specifications, manuals or documented information about customers and suppliers. They are 'visible' and easily digitized into IT, and can be traded and transferred with relative ease.

Routine knowledge assets. These are the tacit knowledge that is routinized and embedded in the actions and practices of the organization.

Know-how, organizational routines, and organizational culture in carrying out the daily business of the organization are examples of such knowledge assets. Sharing narratives and stories about their own company also helps build routine knowledge assets (Nonaka, Toyama, and Nagata, 2000: 15–17; cf. also Nonaka, Toyama, and Konno, 2000; Kokuryo, Nonaka, and Kataoka, 2003; Umemoto, 2002)

Figure 3.4 gives an overview of the four categories of knowledge assets. Since knowledge assets are both inputs and outputs of the organization's knowledge-creating activities, they are constantly evolving (Nonaka and Toyama, 2002: 997). The most important knowledge assets are the capability to continuously create new knowledge out of existing firm-specific capabilities, rather than the stock of knowledge, such as particular technology, that a firm possesses at one point in time (cf., for example, Barney, 1991; Lei, Hitt, and Bettis, 1996; Nelson, 1991; Nonaka and Toyama, 2002; Teece, Pisano, and Shuen, 1997).

According to Nonaka and Toyama (2002: 998), '[h]igh-quality tacit knowledge is the source of sustainable competitive advantage since it takes times to be accumulated and is not easily replicated'. Therefore, 'organizational knowledge, learning and capabilities form a triangle: the ongoing development of organizational knowledge is, or can be, a dynamic capability that leads to continuous organizational learning and further development of knowledge assets' (Tsoukas and Mylonopoulos, 2004: S2).

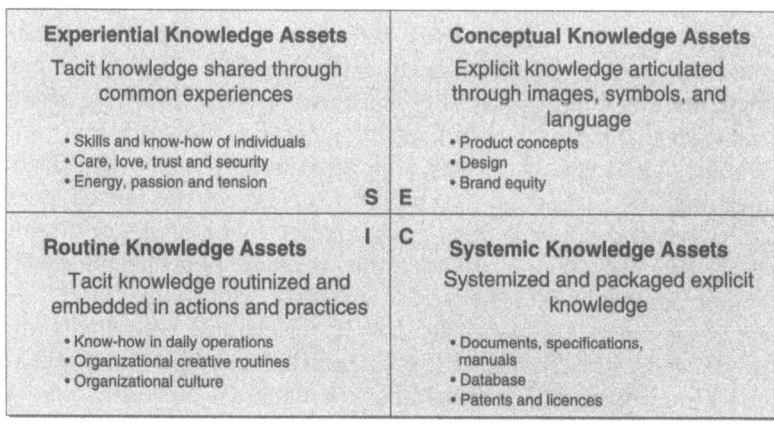

Figure 3.4 Four categories of knowledge assets (from Nonaka, Toyama, and Konno, 2000: 20)

3.5   Communities of practice

Chapter 3.4.1 has shown that certain enabling conditions are necessary for successful organizational knowledge creation (cf. also Ichijo, 2006b; von Krogh, Ichijo, and Nonaka, 2000), and one of the most important ones has turned out to be a shared context or *ba*. As will be discussed below, Western scholars have also come up with a concept similar to *ba*, namely so-called communities of practice (CoP). In fact, a considerable body of research focuses on learning and knowledge sharing in CoPs, a field that has been developed and significantly shaped by the works of Wenger and fellow researchers (Lave and Wenger, 1991; Wenger, McDermott, and Snyder, 2002; Wenger and Snyder, 2000). CoPs have recently become 'key components in an organizational learning toolkit' (Plaskoff, 2003: 161), and can be seen as 'the cornerstones of knowledge management' (Wenger, 2004: 2). As a result, they have achieved prominence in the context of knowledge management and organizational learning both with scholars and practitioners (cf., for example, Brown and Duguid, 1991, 2001; Saint-Onge and Wallace, 2003; Soekijad, Huis in 't Veld, and Enserink, 2004; Swan, Scarbrough, and Robertson, 2002).

Wenger and Snyder (2000: 139) speak of CoPs as 'a new organizational form' that is emerging that promises to complement existing structures of knowledge management and radically galvanize knowledge sharing, learning and change. CoPs can be defined as 'groups of people who share a concern, a set of problems, or a passion about a topic, and who deepen their knowledge and expertise in this area by interacting on an ongoing basis' (Wenger, McDermott, and Snyder, 2002: 4), or more generally as 'an activity system about which participants share understandings concerning what they are doing and what that means in their lives and for their community' (Lave and Wenger, 1991: 98). Thus, they are united in both action and in the meaning that that action has, both for themselves and for the larger collective, and can be defined by disciplines, by problems, or by situations (Wenger, 2004: 2). 'In brief, they're groups of people informally bound together by shared expertise and passion for a joint enterprise – engineers engaged in deep-water drilling, for example, consultants who specialize in strategic marketing, or frontline managers in charge of check processing at a large commercial bank' (Wenger and Snyder, 2000: 139). Finally, CoPs 'appear to be an effective way for organizations to handle unstructured problems and to share knowledge outside of traditional structural boundaries' and serve as 'a means of developing and maintaining long-term organizational memory' (Lesser and Storck, 2001: 832). As a result, community building 'can be viewed as learning how to learn organizationally' (Plaskoff, 2003: 166). According to Wenger, McDermott, and Snyder (2002: 24), CoPs 'vary widely in both name and style in different organizations'. Another term that can frequently be found from the extant literature and that seems to be even more general than the term CoP is 'knowledge community' (KC) – sometimes also referred to as 'strategic communities' – (cf., for example, Barrett, Cappleman, Shoib, and Walsham, 2004; Garavelli, Gorgoglione, and Scozzi, 2004; Hustad, 1999; Storck and Hill, 2000). However, there does not seem to exist a common definition of the term: Lindkvist (2005), for example, speaks of KC in regard to CoPs and of knowledge collectivity with reference to what he suggests as a 'collectivity-of-practice', and Ryu and fellow researchers (2005) view enterprise information portals (EIP) as KCs, which is obviously too simplistic a view.

According to Lave and Wenger (1991), the sharing of expertise and the creation of new knowledge, often tacit in nature, is a central tenet of a CoP's existence; it exists as a social gathering or technological network. The sharing of tacit knowledge by and through CoPs is by means of storytelling, conversation, coaching, and apprenticeship provided by CoPs (Wenger, McDermott, and Snyder, 2002). As a matter of fact, the sharing of tacit knowledge – socialization – as well as its (partial) transformation into explicit knowledge – externalization – are at the heart of CoPs. This also seems to be in line with Nonaka's theory of knowledge creation and Japanese firms' particular focus on tacit knowledge (cf. 3.4). Besides, as managing existing knowledge alone is simply not enough (Umemoto, 2002), the creation of new knowledge and organizational learning are key as well.

According to Plaskoff (2003: 179), '[c]ommunities provide an enabling context for knowledge creation'. Indeed, organization structures and systems that provide a context that co-ordinates and motivates action are critical elements of the overall knowledge organization (Wenger, McDermott, and Snyder, 2002). As they view *ba* as 'an existential place where participants share their contexts and create new meanings through interactions', Nonaka and Toyama (2003: 7; cf. also Nonaka, von Krogh, and Voelpel, 2006) acknowledge similarities of the concept of *ba* to the concept of CoP, but stress important differences (cf. also 3.4.1).[6] In the end, it is probably safe to say that CoPs are, or at least can constitute and provide, a certain type of *ba*, an enabling context for knowledge creation, sharing, and organizational learning in organizations. Indeed, according to Mavin and Cavaleri (2004: 286), learning is 'embedded in and mediated through particular social and cultural contexts' and such social learning in context enhances the performance and capability of organizations. This kind of social learning in context has been termed 'situated learning' by Lave and Wenger (1991).

In fact, '[a]s organizations grow in size, geographical scope, and complexity, it is increasingly apparent that sponsorship and support of groups such as [CoPs] is a strategy to improve organizational performance' (Lesser and Storck, 2001: 831) and '[s]uccess in global markets depends on communities sharing knowledge across the globe' (Wenger, McDermott, and Snyder, 2002: 7). Therefore, CoPs 'can be particularly useful in helping to build a global organization out of a lot of individual operating companies in separate countries' (Buckman, 2004: 164) – Wenger and fellow researchers speak of 'distributed' CoPs (2002) – and thus foster the sharing of knowledge horizontally and across intra-organizational boundaries. Besides, another benefit of CoPs is that they 'evade the ossifying tendencies of large organizations' (Brown and Duguid, 1991: 50).

3.6 Global knowledge-based management

Bartlett and Ghoshal (2002: 3) state that the world's largest companies are in flux and that '[n]ew pressures have transformed the global competitive game'. 'In a global economy, the boundaries of a firm are not geographically determined' (Leonard, 1998: 216). 'Virtually all business conducted today is global business' (Thomas, 2002: 3); national economies have become increasingly deregulated and have opened up opportunities for international trade and competition so that it has 'become the norm for organizations to compete for market share not only with their national competitors but also with international ones' (Trompenaars and Woolliams, 2004: 27). Besides, in such 'an era of ever faster innovation cycles combined with an increasing convergence of industries ... and intense and global competition, advantages tend to erode quickly' (Ambos and Schlegelmilch, 2005: 23). Furthermore, '[i]n today's hyper-competitive global marketplace it is pivotal for enterprises to manage not only tangible resources but also to exploit intangibles' (Desouza and Evaristo, 2003: 62). At the same time – or maybe even specifically for these reasons – '[t]he last two decades have seen an increase in cooperative activity between firms, particularly between trans-national corporations' (Buckley, Glaister, and Husan,

2002: 113). Indeed, numerous studies have noted and commented on this rise in strategic alliance and international joint venture (IJV) formation (Child, Faulkner, and Tallman, 2005; Contractor and Lorange, 2002; Doz and Hamel, 1998; Dyer, Kale, and Singh, 2001, 2004; Grant and Baden-Fuller, 2004; Inkpen, 2000; Inkpen and Ramaswamy, 2006; Lane, Salk, and Lyles, 2001) and this fact 'is taken as further proof of the unstoppable march of globalization, particularly as a large and growing number of these agreements involve firms of at least two nationalities' (Narula and Hagedoorn, 1999: 283). Besides, the benefits of cross-border alliances, foreign partnerships, and joint ventures (JV) – even between competitors – in general, as well as of (inter-)organizational learning and knowledge transfer and acquisition, have frequently been pointed out and discussed (cf., for example, Child, Faulkner, and Tallman, 2005; Dhanaraj, Lyles, Steensma, and Tihanyi, 2004; Hamel, Doz, and Prahalad, 1989; Inkpen, 1998, 2000, 2002; Inkpen and Currall, 2004; Inkpen and Dinur, 1998; Inkpen and Tsang, 2005; Lane, Salk, and Lyles, 2001; Lyles, 1988; Salk and Simonin, 2003; Simonin, 1999b, 2004; Steensma, Tihanyi, Lyles, and Dhanaraj, 2005).

This view seems to be in line with international business and knowledge management scholars' positing the capability of MNCs to create and efficiently transfer and combine knowledge from different locations around the world to become more and more important as a determinant of competitive advantage and as critical to their success and survival (cf., for example, Asakawa and Lehrer, 2003; Bartlett and Ghoshal, 2002; Chini, 2004; Desouza and Awazu, 2005b; Doz, Santos, and Williamson, 2001; Gupta and Govindarajan, 2000a; Macharzina, Oesterle, and Brodel, 2001; Schulz and Jobe, 2001). This cross-border synergistic process of joint knowledge creation has been termed 'global knowledge creation' and identified as a key process of globalization (Nonaka, 1990b: 82).

Besides, as new knowledge 'provides the basis for organizational renewal and sustainable competitive advantage', '[t]he acquisition of new organizational knowledge is increasingly becoming a managerial priority' and, in the global arena, 'the complexities increase in scope as multinational firms grapple with cross-border knowledge transfers and the challenge of renewing organizational skills in various diverse settings' (Inkpen, 1998: 69). In fact, '[t]acit knowledge, embodied in individual, group and organizational routines, is of critical strategic importance because, unlike explicit knowledge, it is both inimitable and appropriable' (Al-Laham and Amburgey, 2005: 251; cf. also Spender, 1996b).

'For learning to be more than a local affair, knowledge must spread quickly and efficiently throughout the organization', as '[i]deas carry maximum impact when they are shared broadly rather than held in a few hands' (Garvin, 1993: 87). According to Bresman, Birkinshaw, and Nobel (1999: 440), the process of knowledge transfer between business units is an essential aspect of knowledge management, and Tseng (2006: 121) notes that knowledge transfer capability is one of the most important advantages of MNCs and that '[t]hrough the transfer and adaptation of knowledge, subsidiaries of MNCs build and develop their competitiveness over local firms'. Indeed, the management of knowledge flows is especially important for MNCs because they operate in geographically and culturally diverse environments (Schulz and Jobe, 2001). According to Teece (2000b: 37), 'the very essence of a large, integrated firm can be traced in substantial measure to its capacity to facilitate the (internal) exchange and transfer of knowledge assets and services, assisted and protected by administrative processes'. Denrell, Arvidsson, and Zander (2004: 1491) argue that the 'importance of knowledge transfer in multinational companies implies that identifying capabilities and expertise, that is, "knowing who in the organization is good at what," is a major component in knowledge management' and that '[i]f knowledge is to be "managed" and transferred, it is essential that participants in multinational companies know (and agree on) where capabilities reside'.

Since strategically important knowledge is geographically dispersed in the business environment of most global firms (Asakawa and Lehrer, 2003; Teece, 2000b), MNCs can derive great competitive advantage by managing knowledge flows between their subunits with differences between local markets requiring adaptation of products and operations to local conditions (Schulz and Jobe, 2001; Haghirian, 2003). Minbaeva and associates (2003: 587) contend that the competitive advantage that MNCs enjoy is contingent upon their ability to facilitate and manage inter-subsidiary transfer of knowledge, and define knowledge transfer between organizational units as 'a process that covers several stages starting from identifying the knowledge over the actual process of transferring the knowledge to its final utilization by the receiving unit'. Doz, Santos, and Williamson (2001: 219) point to the important fact that MNCs will have to shift from merely being global projectors of knowledge to so-called metanational companies, which means 'exploiting the potential of learning from the world by unlocking and mobilizing knowledge that is imprisoned in local pockets scattered around the globe'. However, while leveraging locally-embedded knowledge assets for global use is indeed a major challenge for multinational knowledge management, innovation by local units can also be leveraged for regional application (Asakawa and Lehrer, 2003).

Schulz (2001: 663) defines knowledge flows as 'the aggregate volume of know-how and information transmitted per unit of time' and states that with this definition he intends 'to capture the overall amount of know-how and information transmitted between subunits in all kinds of ways, including via telephone, e-mail, regular mail, policy revisions, meetings, shared technologies, and reviews of prototypes'. He further distinguishes between three subunit learning processes, namely collecting new knowledge, codifying knowledge, and combining old knowledge, with collecting new knowledge occurring, 'when a subunit is exposed to complex, uncharted domains of activity or to environments characterized by a high rate of innovation and change' (Schulz, 2001: 663). Gupta and Govindarajan (1991: 773) – who describe MNCs as a network of capital, product, and knowledge transactions among units in different countries, a perspective which is also consistent with the analyses of Bartlett and Ghoshal (2002) – use the term 'intracorporate knowledge flow' and define it as 'the transfer of either expertise (e.g., skills and capabilities) or external market data of strategic value'. In a further study, they were able to show that a complete mapping of the knowledge transfer process within MNCs requires attention to all of the following five major elements: value of the knowledge possessed by the source unit; motivational disposition of the source unit regarding the sharing of its knowledge; the existence, quality, and cost of transmission channels; motivational disposition of the target unit regarding acceptance of incoming knowledge; and the target unit's absorptive capacity for the incoming knowledge (Gupta and Govindarajan, 2000a). In particular, 'the context specificity of the knowledge has an effect on the extent of knowledge transfer, both because the more context specific the knowledge is, the smaller the absorptive capacity of the received and the less it can be used in other MNC units' (Foss and Pedersen, 2002: 64).

Minbaeva and colleagues' (2003) most important finding, for instance, is that both aspects of absorptive capacity (ability and motivation) need to be present in order to optimally facilitate the absorption of knowledge from other parts of the MNC and that employee ability or motivation alone does not lead to knowledge transfer. Contrary to studies that blame primarily motivational factors, Szulanski's (1996) findings on internal stickiness, in turn, show the major barriers to internal knowledge transfer to be knowledge-related factors such as the recipient's lack of absorptive capacity, causal ambiguity, and an arduous relationship between the source and the recipient (cf. also Szulanski, 2003; Szulanski and Cappetta, 2003; von Hippel, 1994). In fact, whether or not the evaluation of the knowledge results in its integration in the organizational knowledge base depends on the learning effectiveness or absorptive capacity[7] of the organization. Inkpen (1998, 2000) describes three factors influencing learning effectiveness – knowledge connections (such as foreign assignments or visits by personnel) between the partner firms to build networks; relatedness of partner knowledge; and the cultural alignment between parent executives and alliance managers (cf. also 3.7).

Moreover, knowledge is 'simultaneously highly sophisticated (both tacit and explicit) and widely dispersed in the hands and minds of many, and is not easily produced or captured inside the boundaries of one or a few firms' (Ciborra and Andreu, 2001: 78). As mentioned above, Nonaka (1990b: 82) terms the cross-border synergistic process of joint knowledge creation 'global knowledge creation' and sees it as the key process of globalization. Here again, '[t]acit knowledge, embodied in individual, group and organizational routines, is of critical strategic importance because, unlike explicit knowledge, it is both inimitable and appropriable' (Al-Laham and Amburgey, 2005: 251; cf. also Spender, 1996b). According to Teece (2000b: 41), 'the conversion of tacit to codified or explicit knowledge assists in knowledge transfer and sharing, thereby possibly helping to make the firm more innovative and more productive' because '[o]nce knowledge is made explicit, it is easier to store, reference, share, transfer, and hence re-deploy'.

Finally, Dixon (2000: 143) summarizes the following two fundamental messages of her work on common knowledge and knowledge transfer: '(1) there are many, very different ways to transfer knowledge, and (2) knowledge is transferred most effectively when the transfer process "fits" the knowledge being transferred'. Figure 3.5 gives an overview of Dixon's model of knowledge transfer based on the type of task.

Obviously, managing knowledge transfers in firms is a very complex and difficult task and so is researching this process. This also explains the vast amount of theories and different research streams in this area. In Chapter 4.2, I will discuss the co-creation[8] and transfer – that is, recreation – of marketing knowledge in MNCs in theory and in Chapters 5 and 6 in relation to the empirical research project.

Type of transfer	Definition	Nature of task	Type of knowledge
Serial transfer	Previous knowledge for an analogous situation	Frequent and non-routine	Tacit and explicit
Near transfer	Previous knowledge for very similar situation	Frequent and routine	Explicit
Far transfer	Non-routine knowledge for similar task	Frequent and non-routine	Tacit
Strategic transfer	Applying collective knowledge (as in a merger)	Infrequent and non-routine	Tacit and explicit
Expert transfer	Acquiring expert knowledge from elsewhere	Infrequent and routine	Explicit

Figure 3.5 Dixon's model of knowledge transfer based on type of task (based on Dixon, 2000: 144–5)

3.7 Inter-organizational knowledge-based management

Learning and knowledge management have become a key alliance research issue in recent years (cf., for example, Desouza and Awazu, 2005b; Inkpen, 2002; Inkpen and Currall, 2004). Since alliances can be defined as 'any inter-firm cooperation that falls between the extremes of discrete, short-term contracts and the complete merger [or acquisition] of two or more organizations' (Contractor and Lorange, 2002: 486), it becomes obvious that concepts from alliance learning research might also provide helpful insights for knowledge and learning issues in acquisitions.

Research on IJVs and theory development has greatly advanced since Parkhe's (1993: 227) pronouncement to the effect that IJVs 'lack a strong theoretical core or an encompassing framework that effectively integrates past research and serves as a springboard for launching future research'. According to Ahmadjian and Lincoln (2001: 684), research on inter-firm alliances has blossomed over the last decade, with there being 'an intellectual tension between two dominant approaches to alliance – governance and learning'. This chapter focuses on the latter.

3.7.1 Knowledge creation, transfer, and organizational learning in IJVs

In recent years, learning and knowledge management have become a key alliance research issue (Desouza and Awazu, 2005b; Inkpen, 2002; Inkpen and Currall, 2004). In fact, '[m]any alliances are established in order to enhance a company's knowledge or capacity to generate new knowledge through learning' and as strategic alliances – including JVs, collaborations, and consortia – 'are at base all about organizational learning' they should be structured towards that end (Child, Faulkner, and Tallman, 2005: 271, 7). Therefore, the benefits of cross-border alliances, foreign partnerships, and JVs – even between competitors – have frequently been discussed, often with a focus on (inter-)organizational learning and knowledge transfer and acquisition (cf., for example, Chaston, 2004; Child, Faulkner, and Tallman, 2005; Desouza and Awazu, 2005b; Dhanaraj, Lyles, Steensma, and Tihanyi, 2004; Hamel, Doz, and Prahalad, 1989; Inkpen, 1998, 2000, 2002; Inkpen and Currall, 2004; Inkpen and Dinur, 1998; Inkpen and Tsang, 2005; Lane, Salk, and Lyles, 2001; Lyles, 1988; Salk and Simonin, 2003; Simonin, 1999b, 2004; Steensma, Tihanyi, Lyles, and Dhanaraj, 2005). Put in a nutshell, IJVs are viewed as effective conduits that enable MNCs to exploit their knowledge in multiple markets (Dhanaraj, Lyles, Steensma, and Tihanyi, 2004) and learning – together with trust and control – has become one of the most important and studied concepts in the alliance and JV literatures (Inkpen and Currall, 2004). Indeed, 'since not all critical knowledge resides inside firm boundaries, firms have to tap into external resources of knowledge to develop competitive advantage' (Al-Laham and Amburgey, 2005: 251; cf. also Cavusgil, Calantone, and Zhao, 2003; Desouza and Awazu, 2005b). Obviously, IJVs and other kinds of alliances are a case in point here as they have often been considered a central source of new knowledge (Gulati, Nohria, and Zaheer, 2000; Hamel, 1991; Khanna, Gulati, and Nohria, 1998; Kogut, 1988; Lyles, 1994) and access to the capabilities of the partners has been emphasized as a central motive for such 'learning alliances' (Badaracco, 1991; Child, Faulkner, and Tallman, 2005; Lane, Salk, and Lyles, 2001; Lubatkin, Florin, and Lane, 2001; Mowery, Oxley, and Silverman, 1996). Indeed, as Iansiti and Levien (2004a: 1) have postulated, '[s]trategy is becoming, to an increasing extent, the art of managing assets that one does not own'.

According to Inkpen and Currall (2004: 586), there are 'various types of strategic alliances, such as joint ventures, licensing agreements, distribution and supply agreements, research and development partnerships, and technical exchanges' (cf. also Contractor and Lorange, 2002; Grant and Baden-Fuller, 2004; Gulati and Singh, 1998; Inkpen, 1998). Following Inkpen and Currall (2004: 586), I also focus on 'equity joint ventures', 'an alliance form that combines resources from more than one organization to create a new organizational entity (the "child") distinct from its parents'. In the case of IJVs, two or more parent companies from at least two nations establish a jointly owned entity in which the operational management is often shared by on-site representatives from both parents (Osland and Cavusgil, 1998: 192). As a matter of interest, Mowery and fellow researchers (1996) have shown that equity JVs are more effective for the acquisition of knowledge associated with partner capabilities than contract-based alliances (cf. also Inkpen, 1998).

Inkpen (1998: 72) uses the term 'alliance knowledge' to indicate 'knowledge from an alliance [that] can be used by the parent company to enhance its own strategy and operations' and contrasts it with knowledge 'about how to design and manage alliances' (Inkpen, 1998: 71) – an issue dealt with, for example, by Lyles (1988) – and the situation where 'parent firms may seek collaborative access to other firms' knowledge but will not necessarily wish to internalize the knowledge in their own operations' (Inkpen, 1998: 71–2).

Indeed, Inkpen (1998, 2000) identifies three main conditions that enable the exploitation of learning opportunities provided by alliances: value, accessibility, and learning effectiveness.

Value. In order to enter into the process of knowledge sharing, high value must be attached to alliance knowledge because '[a]n alliance partner's approach to knowledge acquisition will be a function of the perceived value of alliance knowledge' (Inkpen, 1998: 72). That means that the value attached to the knowledge stored in a partner organization must be higher than the expected cost of the knowledge acquisition. Besides, organizational units filter information according to their (culturally influenced) systems of meaning and funds of knowledge and subsequently tend to ignore information that is of low relevance to the local task but that might be of high importance to the global task: 'what counts as valuable knowledge does not appear to be fixed but rather derives at least in part from social conventions that differ from one social context to the next' (Macharzina, Oesterle, and Brodel, 2001: 636).

Accessibility. Alliance knowledge must be accessible to the partners but there are two factors limiting knowledge accessibility: partner protectiveness and knowledge tacitness (Inkpen, 1998, 2000). Indeed, for competitive reasons, alliance partners may be highly protective of their knowledge resources. But increasing trust between alliance partners may mitigate partner protectiveness. Besides, partners may decrease their efforts to protect knowledge spillovers over time and alliance knowledge will become more accessible, specifically as trust increases and mutual partner understanding develops (Inkpen, 1998; Inkpen and Currall, 2004). DeLong and Fahey (2000: 119) put it like this: 'The level of trust that exists between the organization, its subunits, and its employees greatly influences the amount of knowledge that flows both between individuals and from individuals into the firm's databases, best practices archives, and other records' (cf. also Davenport and Prusak, 2000; Child, Faulkner, and Tallman, 2005; Inkpen and Currall, 2004; Madhok, 2006, specifically in alliance contexts). In fact, only in a climate of trust will organizations be ready to put their knowledge at the disposal of their partner organizations, as '[w]e are now challenged to create organizations based on cultures of trust that will support the dynamic teaming of capable individuals and companies and add to the knowledge creation process (Savage, 1996)' (Bertels and Savage, 1999: 208).

According to Inkpen (1998: 74), '[o]rganizational knowledge creation involves a continuous interplay between tacit and explicit knowledge' (cf. also Nonaka and Takeuchi, 1995; von Krogh, Ichijo, and Nonaka, 2000). Where knowledge creation and sharing in an interorganizational context are concerned, organizational structures should reinforce tacit–explicit knowledge interaction across many different boundaries (Ichijo, 2006b). However, '[t]he more tacit the knowledge that an alliance partner seeks to acquire, the more difficult the acquisition', while at the same time the likelihood that the knowledge is valuable rises with the tacitness of the knowledge (Inkpen, 1998: 74).

Learning effectiveness. Accessibility is not sufficient for effective learning and the partners' effectiveness at learning and acquiring knowledge is important as well. Learning effectiveness is closely related to the concept of 'absorptive capacity' (Inkpen, 1998; cf. also Cohen and Levinthal, 1990; Van den Bosch, Van Wijk, and Volberda, 2003). Inkpen (1998: 75) found that three factors influence learning effectiveness in the alliance context: knowledge connections between a firm and its alliance; the relatedness of alliance knowledge; and the cultural alignment between parent executives and alliance managers. 'Knowledge connections occur through both formal and informal relationships between individuals and groups and can be seen as internal managerial relationships that facilitate the sharing and communicating of new knowledge and provide a basis for transforming individual knowledge to organizational knowledge' (Inkpen, 1998: 75).

Knowledge connections tend to evolve and intensify over time. 'Prior knowledge permits the effective utilization of new knowledge. New knowledge in an area we are familiar with is generally easier to acquire than knowledge about an unfamiliar area' (Inkpen, 1998: 76). There are two types of related knowledge important in this context: knowledge of the partner and knowledge about alliance management. Generally, 'the greater the difference between the partner firms, the more difficult it is to create a learning relationship, and the greater the probable value of learning' (Inkpen, 1998: 76). Indeed, '[e]thnocentrism, skepticism of the credibility of remote sources, suspicion of the unknown, and resistance to change can lead organizational units to reject proposals' (Macharzina, Oesterle, and Brodel, 2001: 647).

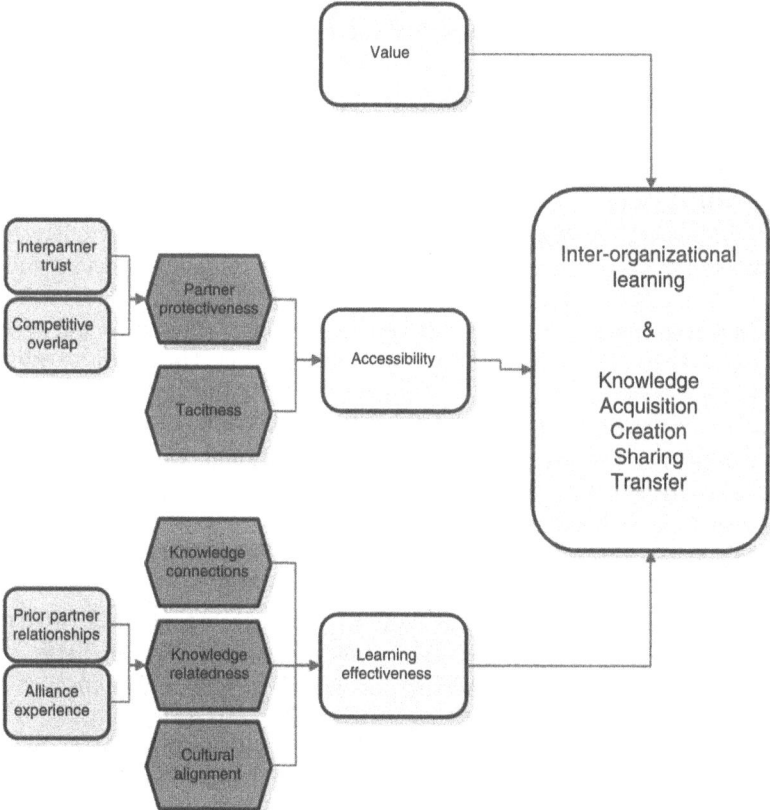

Figure 3.6 Inkpen's knowledge acquisition framework (author's own illustration based on Inkpen, 1998, 2000)

'Cultural alignment' derives from a term from Schein (1996) and refers mainly to the degree of difference of assumptions about the alliance relationship, objectives and performance (Inkpen, 1998). Figure 3.6 summarizes the above framework.

Last but not least, Ahmadjian (2004) notes that inter-organizational knowledge creation requires a *ba* (3.4), or a space for interaction to encourage the inter-organizational community to engage in the spiralling knowledge-creating process (cf. Chapter 3.4.1). In the process of inter-organizational knowledge creation, organizations must also find ways to create the same sort of context or *ba* between firms, that is, 'ways must be found to nurture a culture, a language to facilitate exchange of ideas, and an atmosphere of trust and care' (Ahmadjian, 2004: 229, 230).

3.7.2 A theoretical framework of learning and knowledge creation in IJVs

As we have seen, there are usually three entities involved in an IJV: the two partner firms (parents) and the IJV itself (child). Therefore, learning, creation, and transfer of alliance knowledge can take place on three different levels (cf. Figure 3.7): (1) between the two parents, (2) between each parent and the child, and (3) at the IJV. Obviously, in the latter case, the created knowledge and lessons learned can subsequently be transferred to the parents, as in (2). If managers and other employees return or are transferred to the parents, learning between the two parents can also occur on an indirect level. The direct learning between parents, as in (1), might be because of an overall increase in co-operation or at least in communication and contact in the course of establishing and maintaining the IJV, even though this learning and knowledge creation process need not necessarily be directly related to issues concerning the IJV.

Most learning and knowledge-related research on IJVs has focused on the transfer or the acquisition of knowledge – mostly – through parent firms but the critical issue of (co-)creating new knowledge through collaboration has hardly been touched. Indeed, Inkpen's (1998, 2000) knowledge acquisition framework, as outlined above, offers valuable insights into the process of learning through JVs, but knowledge creation – though often briefly discussed (most notably by Inkpen and Dinur, 1998) – is not analysed in detail. This chapter leads to two fundamental conclusions: first, mutual learning and knowledge (co-)creation in IJVs is a crucial issue and essential to gaining and sustaining competitive advantage for the IJV (child), and in the optimum

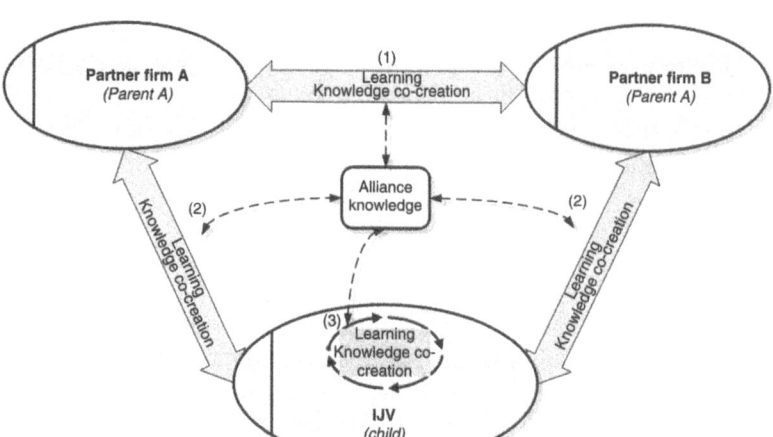

Figure 3.7 Inter-organizational learning and knowledge creation on three different levels (author's own illustration)

case also for both partner firms (parents); second, establishing strategic alliances such as IJVs can also be seen as a helpful means to source external knowledge by tapping into the expertise and knowledge base of other firms, as well as to co-create new knowledge through co-operation and interaction with a partner. However, mutual learning and knowledge co-creation in and through IJVs is by no means an easy and straightforward task. The influencing factors identified by Inkpen (see above) and the enabling conditions identified by Ichijo (2006b) play a crucial role in making the endeavour succeed or fail.

3.8 Conclusion

Dorothy Leonard's[9] work (Leonard, 1998; Leonard and Rayport, 1997; Leonard and Sensiper, 1998; Leonard, 2000; Leonard and Swap, 2004, 2005a, 2005b; Leonard-Barton, 1992) has also centred around knowledge assets and their management in organizations. She is one of the most cited scholars in the field of knowledge-based management and innovation (cf., for example, Choo, 2003; Takeuchi and Nonaka, 2004) and her book *Wellsprings of Knowledge* (1998) – which originally appeared in 1995 – is a classic. It starts with the following introduction:

> This book is about a process that sounds abstract and yet is concrete, practical, and profoundly important – managing a firm's knowledge

assets. Companies, like individuals, compete on the basis of their ability to create and utilize knowledge; therefore, managing knowledge is as important as managing the finances. In other words, firms are knowledge, as well as financial, institutions. They are repositories and wellsprings of knowledge. (Leonard, 1998: xi)

The 'starting point for managing knowledge in an organization is an understanding of core capabilities', which 'constitute a competitive advantage for a firm' and 'have been built up over time and cannot be easily imitated' (Leonard, 1998: 4). In fact, rather than 'static or publicly available, capabilities are largely tacit and have to be acquired in an idiosyncratic and path-dependent way via social learning by doing and imitation' (Taudes, Trcka, and Lukanowicz, 2002: 142). It is exactly these core competencies that are building and sustaining the sources of innovation in organizations. They are created through knowledge-creating activities, but those activities are also dependent on, and enabled by, core capabilities (Leonard, 1998: 4–5). On the other hand, '[v]alues, skills, managerial systems, and technical systems that served the company well in the past and may still be wholly appropriate for some projects or parts of projects, are experienced by others as core rigidities – inappropriate sets of knowledge', the flip side of core capabilities, so to speak (Leonard-Barton, 1992: 118; cf. also Leonard, 1998). They are 'activated when companies fall prey to insularity or overshoot an optimal level of best practices' (Leonard, 1998: 55). Finally, Leonard (1998: 266–7) concludes the book with the following:

> Wellsprings of knowledge not only feed the corporation but are fed from many sources. If all employees conceive of their organizations as a knowledge institution and care about nurturing it, they will continuously contribute to the capabilities that sustain it.

In reviewing the literature on the knowledge-based view of the firm, striking similarities between the theories of two of the most prominent and eminent scholars in the field of knowledge management – Dorothy Leonard and Ikujiro Nonaka – become obvious.[10] Indeed, Nonaka's theory of organizational knowledge creation (cf. 3.4) and Leonard's work on knowledge assets seem to be strongly related and interconnected. The most striking similarities or common foci are:

- new product development;
- tacit knowledge;

- knowledge creation;
- innovation.

Both Leonard and Nonaka focus their research on new product development projects, emphasize the importance of tacit knowledge and
knowledge creation – rather than merely managing it – and link the
knowledge-creating process to the generation of innovations. Given
this significant coherence of the two theories, as well as their impact
and the fact that they are widely received and acclaimed, make them
appear suitable as the building blocks of the theoretical framework of
this book.[11] Moreover, as new product development and innovation
are also closely related to the field of marketing, the two theories will
have important implications for developing a knowledge-based theory
of marketing.

To conclude, let us consider the following quote from Teece
(2000b: 51):

> Today's competitive environment favours organizations (firms) able
> to protect knowledge assets from re-contracting hazards, but it also
> favours firms which can build, buy, combine, deploy and re-deploy
> knowledge assets according to changing customer needs and the
> changing competitive circumstances. Successful firms of the future
> will be 'high flex' and knowledge-based.

This statement clearly emphasizes the importance of knowledge assets
and the need for a knowledge-based approach to management.

4

Knowledge-based Management and Organizational Learning in Marketing

> We must keep in mind that the core of business is production and marketing/sales. (Gummesson, 2003a: 483)

From a knowledge-based perspective (cf. also Chapter 3), business organizations are viewed as bundles or as a collection of knowledge assets, the effective management of which affords firms competitive advantage (Choo and Bontis, 2002a; Nonaka and Toyama, 2002; Tsoukas and Mylonopoulos, 2004; cf. also Day, 1994a, in the field of marketing). Therefore, Teece (2000a) argues that the core of management in the knowledge economy is the development and deployment and utilization of intangible assets, the most significant being (tacit) knowledge, competence, and intellectual property. In this chapter, this knowledge-based view of management is applied to the field of marketing.

This section explores the extant literature on marketing knowledge, its creation and management as well as on organizational learning in marketing functions. It also introduces and defines important key concepts. As will be shown, different streams of research have contributed to the field but all in all it may not be outlandish to conclude that research on knowledge-based approaches to marketing is still rather in its infancy. As mentioned above, one aim of this book is to develop a conceptual framework and model of knowledge-based marketing. To do so, at least two essential questions have to be answered (cf. also 2.2): (1) What is marketing knowledge? and (2) What is its role in marketing and how is it created and managed?

4.1 Theoretical background: literature review and state-of-the-field[12]

Tzokas and Saren (2004: 125) contend that in marketing, 'knowledge constitutes the basic tenet of the marketing concept as this is expressed by means of market orientation'. Therefore, before setting out to answer the above questions, I shall discuss this important antecedent of knowledge-based marketing, namely the concept of market orientation, which has also frequently been combined with organizational learning.

4.1.1 Market orientation and organizational learning

In the late 1980s and the beginning of the 1990s a research stream on 'market orientation' evolved (for example, Day, 1990; Day, 1994a; Deshpandé and Webster, 1989; Kohli and Jaworski, 1990; Kohli, Jaworski, and Kumar, 1993; Narver and Slater, 1990; cf. Deshpandé, 1999, for a collection of the most influential articles) that focuses on the processes underlying the ability of organizations to generate knowledge about customers, competitors and other players (Vicari and Cillo, 2006). Indeed, the more recent interest in and emphasis on market orientation by and large relates to acquisition and exploitation of knowledge about customers and competitors (Grønhaug, 2002; Slater and Narver, 1995). Market orientation 'reflects a [firm's] ability to internalize the marketing concept as a primary organizing principle of the firm' (Baker and Sinkula, 2005: 483) and 'has emerged as an important area within marketing' (Ottesen and Grønhaug, 2004b: 521) in the meantime (cf., for example, Baker and Sinkula, 2005; Chakravarty, 2000; Day, 1999a; Deshpandé, 1999; Hult, Ketchen, and Slater, 2005; Kirca, Jayachandran, and Bearden, 2004; Menguc and Auh, 2006; Narver, Slater, and MacLachlan, 2004; Ottesen and Grønhaug, 2002, 2004a; Singh, 2004; Webster, 2002). Finally, a significant body of research illustrating the relationship between market orientation and performance has emerged as well[13] (for example, Dawes, 2000; Day, 1999a; Deshpandé, Farley, and Webster, 1993; Deshpandé and Farley, 2004; Hult, Ketchen, and Slater, 2005; Jaworski and Kohli, 1993; Langerak, 2003; Narver and Slater, 1990; Narver, Slater, and MacLachlan, 2004; Ruekert, 1992; Slater and Narver, 1994; cf. also the references in these publications as well as the overview and references in Vicari and Cillo, 2006).

4.1.1.1 Defining market orientation

In his article 'What the hell is "market oriented"?', Shapiro (1988) identified three characteristics that make a company market driven:

(1) *Information on all important buying influences permeates every corporate function.* This is because a company can be market oriented 'only if it completely understands its markets and the people who decide whether to buy its products and services' (Shapiro, 1988: 120). (2) *Strategic and tactical decisions are made interfunctionally and interdivisionally.* In order to 'make wise decisions, functions and units must recognize their differences' and a 'big part of being market driven is the way different jurisdictions deal with one another' (Shapiro, 1988: 122). (3) *Divisions and functions make well-coordinated decisions and execute them with a sense of commitment.* 'An open dialogue on strategic and tactical trade-offs is the best way to engender commitment to meet goals' and '[p]owerful internal connections make communication clear, coordination strong, and commitment high' (Shapiro, 1988: 122).

In the relevant literature, the terms 'market oriented', 'market driven', and 'customer focused' have basically been used interchangeably. By considering these terms to be synonymous, I follow Shapiro (1988), Deshpandé and Webster (1989), Deshpandé, Farley, and Webster (1993), and Slater and Narver (1995) and others.[14] According to Ottesen and Grønhaug (2004b: 521), in the research literature on market orientation, 'the market orientation construct is central' and '[s]everal attempts have been made to delineate and clarify the specific content of this particular concept' (for example, Deshpandé and Farley, 1998; Gray, Matear, Boshoff, and Matheson, 1998; Kohli and Jaworski, 1990; Matsuno, Mentzer, and Rentz, 2000; Narver and Slater, 1990). However, 'an agreed on definition of market orientation does not appear to exist' (Ottesen and Grønhaug, 2004b: 521).

The literature review revealed that the research by Jaworski and Kohli (4.1.1.1.1) and Narver and Slater (4.1.1.1.2) are both the pioneering as well as the most influential works on market orientation. At the same time, Day's work on market-driven organizations (4.1.1.1.3) amended and complemented the two bodies of research and has become recognized and widely cited as well. The following sections (4.1.1.1.1, 4.1.1.1.2, 4.1.1.1.3, and 4.1.1.1.4) will therefore briefly introduce and summarize these three prominent and widely accepted approaches to the concept of market orientation by drawing also from the work of other scholars in the field.

4.1.1.1.1 Jaworski and Kohli.   Jaworski and Kohli's two articles in 1990 and 1993 reinforced the debate on market orientation and triggered a new discussion in the 1990s. In the first article, they argue that even though 'the marketing concept is a cornerstone of the marketing discipline, very little attention has been given to its implementation', and they therefore use the term 'market orientation' 'to mean the implementation of the marketing concept', that is, 'a market-oriented organization is one whose actions are consistent with the marketing concept' (Kohli and Jaworski, 1990: 1). Having identified a lack of clear definition of the concept despite its widely acknowledged importance, the authors synthesize extant knowledge on the subject and provide a foundation for future research by clarifying the construct's domain, developing research propositions, and constructing an integrating framework that includes antecedents and consequences of a market orientation. For them, the starting point of a market orientation is market intelligence and consists of three elements: intelligence generation, intelligence dissemination, and responsiveness (Kohli and Jaworski, 1990: 4–6). Their definition of market orientation therefore reads: 'Market orientation is the organizationwide *generation* of market intelligence pertaining to current and future customer needs, *dissemination* of the intelligence across departments, and organizationwide *responsiveness* to it' (Kohli and Jaworski, 1990: 6, original emphasis). Furthermore, they suggest that firms can gain knowledge or generate market intelligence through the use of market research and they relate this type of research not only to market, sales, pricing, promotion, and customers, but also environmental scanning (Kohli and Jaworski, 1990). In their second article, Jaworski and Kohli (1993) address the following three questions: (1) Why are some organizations more market-oriented than others? (2) What effect does a market orientation have on employees and business performance? (3) Does the linkage between a market orientation and business performance depend on the environmental context? The findings from two national samples suggest that a market orientation is related to top management emphasis on the orientation, risk aversion of top managers, interdepartmental conflict and connectedness, centralization, and reward system orientation. Furthermore, the findings suggest that a market orientation is related to overall (judgmental) business performance (but not market share), employees' organizational commitment, and *esprit de corps*. Finally, the linkage between a market orientation and performance appears to be robust across environmental contexts that are characterized by varying degrees of market turbulence, competitive intensity, and technological turbulence (Jaworski and Kohli, 1993).

4.1.1.1.2 Narver and Slater. Drawing from the extant literature, Narver and Slater (1990: 21) define market orientation as the 'organizational culture ... that most effectively and efficiently creates the necessary behaviors for the creation of superior value for buyers and, thus, continuous superior performance for the business'. More specifically, market orientation is 'the culture that (1) places the highest priority on the profitable creation and maintenance of superior customer value while considering the interests of other key stakeholders; and (2) provides norms for behavior regarding the organizational development of and responsiveness to market information' (Slater and Narver, 1995: 67). Slater and Narver (1995: 63) contend that a market orientation is 'valuable because it focuses the organization on (1) continuously collecting information about target-customers' needs and competitors' capabilities and (2) using this information to create continuously superior customer value'.

Narver and Slater's (1990) notion of market orientation is a one-dimensional construct that comprises three different behavioural components – customer orientation, competitor orientation, and interfunctional co-ordination – and two decision criteria: long-term focus and profitability (cf. also Hult, Ketchen, and Slater, 2005; Vicari and Cillo, 2006). Customer orientation and competitor orientation 'include all of the activities involved in acquiring information about the buyers and competitors in the target market and disseminating it throughout the business(es)' and interfunctional coordination 'is based on the customer and competitor information and comprises the business's co-ordinated efforts, typically involving more than the marketing department, to create superior value for the buyers' (Narver and Slater, 1990: 21). In sum, the three behavioural components of a market orientation comprehend the activities of market information acquisition and dissemination and the co-ordinated creation of customer value (ibid.). Narver and Slater's (1990) inferences about the behavioural content of market orientation are therefore consistent with the findings of Kohli and Jaworski (1990). Their study is an important first step in validating the market orientation–performance relationship, even though the generalizability of the findings is limited.

More will be said on Narver and Slater's approach in section 4.1.1.2.

4.1.1.1.3 Day. Much of Day's work has focused on or is related to market orientation and market-driven organizations (for example, Day, 1990, 1994a, 1994b, 1998, 1999a, 1999b, 2003; Day and Nedungadi, 1994; Day and Schoemaker, 2006; Day and Wensley, 1988). His 1990 book *Market Driven Strategy* can be seen as the starting point or at least his first comprehensive account of this topic.[15] There he contends that '[a]t the heart of a market-driven organization is a deep and enduring commitment to a philosophy that the customer comes first, embodying Drucker's dictum that the purpose of a business is to attract and satisfy customers at a profit'[16] (Day, 1990: 356). But he also reminds us that being customer-oriented is only a necessary, but not sufficient condition and that 'market-driven organizations must meet a dual standard: keep close to the customer, and ahead of competition' (ibid.). Summarizing, a market-driven organization has: (1) commitment to a set of processes, beliefs, and values that permeate all aspects and activities, that are (2) guided by a deep and shared understanding of customers' needs and behaviour, and competitors' capabilities and intentions, for the purpose of (3) achieving superior performance by satisfying customers better than the competitors (Day, 1990: 358).

In his 1999 sequel *The Market Driven Organization*, Day (1999a: ix) proffers that 'in an era of increasing market turbulence and intensifying competition, a robust market orientation has become a strategic necessity'. His refined definition of a market-driven firm is 'a superior ability to understand, attract and keep valuable customers' (Day, 1999a: 5) and he identifies three elements of a market orientation: (1) an *externally oriented culture* with the dominant beliefs, values, and behaviours emphasizing superior customer value and the continual quest for new sources of advantage; (2) *distinctive capabilities* in market sensing, market relating, and anticipatory strategic thinking; and (3) a *configuration* that enables the entire organization continually to anticipate and respond to changing customer requirements and market conditions (Day, 1999a: 6–7). Supporting these three elements is a *shared knowledge base* in which the organization collects and disseminates its market insights, and this knowledge builds relationships with customers, informs the company's strategy and increases the focus of employees on the needs of the market (Day, 1999a: 7). Note that '[c]apabilities are further obscured because much of their knowledge component is tacit and dispersed' (Day, 1994a: 39).

Finally, market-driven organizations have superior market sensing, customer linking, and channel bonding capabilities (Day, 1994a). In fact, the 'ability of the firm to learn about customers, competitors, and channel members in order to continuously sense and act on events and trends in present and prospective markets' is critically important (Day, 1994a: 43; cf. also Day, 1994b). Market-driven firms 'stand out in their ability to continuously sense and act on events and trends in their markets' (Day, 1994b: 9) and 'are better equipped to make fact-based decisions, because they can make market knowledge available to the entire organization' (Day and Montgomery, 1999: 9). In sum, market orientation for him is a firm-level capability that links a firm to its

external environment and enables the business to compete by anticipating market requirements ahead of competitors and by creating
durable relationships with customers, channel members, and suppliers
(Day, 1994a; cf. also Kyriakopoulos and Moorman, 2004). Finally, a
'market-driven culture supports the value of thorough market intelligence and the necessity of functionally coordinated actions directed at
gaining a competitive advantage' (Day, 1994a: 43).

4.1.1.1.4 Summary. Narver and Slater's (1990) view of market orientation is 'the extent to which culture is devoted to meeting customers'
needs and outwitting competitors' (Hult, Ketchen, and Slater, 2005:
1173), while Kohli and Jaworski's (1990) concept is 'the priority placed
on generating, disseminating, and interpreting information about customer needs' (Hult, Ketchen, and Slater, 2005: 1173; cf. also Sinkula,
1994). This basically means that Narver and Slater (1990) – as well as
Day (1990, 1994a, 1994b), Deshpandé and Webster (1989), and Deshpandé, Farley, and Webster (1993) – describe market orientation as a
form of culture, while Kohli and Jaworski (1990: 1) – who, as will be
recalled, describe it as 'the implementation of the marketing concept' –
offer a behavioural definition (cf. also Deshpandé and Farley, 2004;
Homburg, Workman Jr., and Jensen, 2000; Slater and Narver, 1995).
Finally, in a sense, Day's concept of market-driven organizations can
be seen as reconciling both approaches. Indeed, for him, market orientation represents superior skills in understanding and satisfying customers and its principal features are the following (Day, 1990, 1994a):
(1) a set of beliefs that puts the customer's interest first (Deshpandé,
Farley, and Webster, 1993); (2) the ability of the organization to generate, disseminate, and use superior information about customers
and competitors (Kohli and Jaworski, 1990); and (3) the co-ordinated
application of interfunctional resources to the creation of superior
customer value (Narver and Slater, 1990; Shapiro, 1988).

Last, but not least, Kohli and Jaworski (1990), Day (1994a, 1994b),
and Sinkula (1994) argue that market orientation, as an overall
organizational value system, provides strong norms for sharing of
information and reaching a consensus on its meaning (cf. also Slater
and Narver, 1995). Indeed, the market orientation philosophy 'generally means learning about market developments, sharing this information with appropriate personnel, and adapting offerings to a
changing market' (Jaworski, Kohli, and Sahay, 2000: 45). A strong
market orientation manifests itself through customer-focused market-
oriented learning (cf., e.g. Day, 1994b; Jaworski and Kohli, 1993; Kohli

and Jaworski, 1990; Narver and Slater, 1990; Slater and Narver, 1995), and firms with strong market orientations 'prioritize learning about (1) customers (e.g., likes and dislikes, satisfaction, perceptions); (2) factors that influence customers (e.g., competition, the economy, sociocultural trends); and (3) factors that affect the ability of the firm to influence and satisfy customers (e.g., technology, regulation)' (Baker and Sinkula, 2005: 483). Homburg and Pflesser (2000) developed a multiple-layer model of market-oriented organizational culture and – among other results – found that a market-oriented culture appears especially important in a turbulent market environment (cf. also Deshpandé and Farley, 2004).

Summarizing the literature, Kyriakopoulos and Moorman (2004: 223–4) view market orientation as (cf. also Figure 4.1): (1) a firm-level belief or unifying frame of reference that emphasizes serving the customer (Deshpandé, Farley, and Webster, 1993; Homburg and Pflesser, 2000) or understanding buyers' current and latent needs so as to create value for them (Narver and Slater, 1990; Slater and Narver, 1999); (2) a set of organization-wide processes involving the generation, dissemination, and responsiveness to intelligence pertaining to current and future customer needs (for example, Jaworski and Kohli, 1993; Kohli and Jaworski, 1990; Kohli, Jaworski, and Kumar, 1993); and (3) a firm-level capability that links a firm to its external environment and enables the business to compete by anticipating market requirements ahead of competitors and by creating durable relationships with customers, channel members, and suppliers (Day, 1994a).

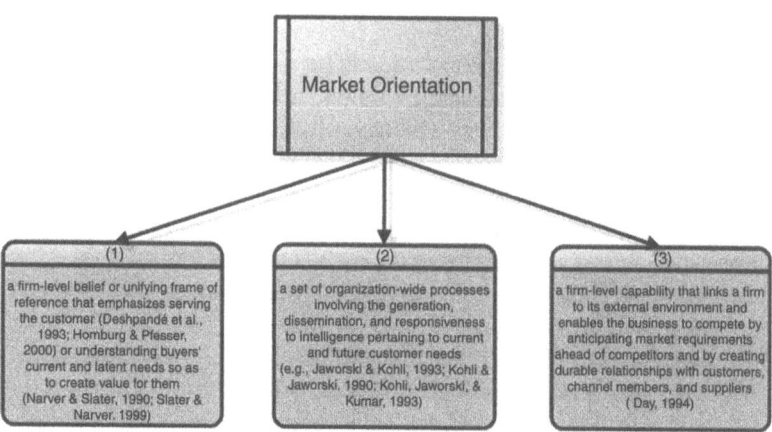

Figure 4.1 Conceptualizations of market orientation (author's own illustration based on Kyriakopoulos and Moorman, 2004)

Finally, it can be said that there are basically two purposes of the extant studies on market orientation. First, they try to identify the activities and processes of an organization that describe its market orientation, and second, they seek to analyse the relationship between an organizational market orientation and an organization's innovativeness (Vicari and Cillo, 2006: 188; cf. also Han, Kim, and Srivastava, 1998; Hurley and Hult, 1998; Li and Calantone, 1998; Slater and Narver, 1995). In these studies, the construct of market orientation is defined in terms of both processes and content of the market intelligence process (Vicari and Cillo, 2006). In fact, as mentioned above, Kohli and Jaworski (1990: 6, emphases removed) define market orientation as 'the organizationwide generation of market intelligence pertaining to current and future customer needs, dissemination of the intelligence across departments, and organizationwide responsiveness to it'. Therefore, 'market orientation is an information-based construct, centered not only on customers, but also on competitors and players working in other industries' and there is 'a relevant difference between those firms that are customer-led and those that are market-oriented' (Vicari and Cillo, 2006: 189; cf. also Slater and Narver, 1998, 1999). Indeed, 'market-oriented businesses scan the market more broadly, have a longer-term focus, and are much more likely to be generative learners' (Slater and Narver, 1998: 1003). Market orientation emphasizes market knowledge sharing and use as central processes to enhance organizational innovative performance (Vicari and Cillo, 2006).

4.1.1.2 *Market orientation and organizational learning*

Basically, from the above, the link between market orientation and organizational learning should already have become obvious. Indeed, much of the literature on market orientation explicitly or implicitly refers to and/or draws from the organizational learning and learning organization literature (for example, Baker and Sinkula, 1999a, 1999b; Bell, Whitwell, and Lukas, 2002; Bennet, 1998; Darroch and McNaughton, 2003; Hurley and Hult, 1998; Kyriakopoulos and Moorman, 2004; Morgan, 2004; Morgan, Katsikeas, and Appiah-Adu, 1998; Morgan and Turnell, 2003; Santos-Vijande, Sanzo-Pérez, Álvarez-González, and Vázquez-Casielles, 2005; Sinkula, 1994; Sinkula, Baker, and Noordewier, 1997; Slater and Narver, 1995). According to Bell, Whitwell, and Lukas (2002: 79), market orientation is an important area of application for organizational learning researchers for a number of reasons. First, the organizational learning and market orientation domains are often perceived as conceptually similar, because – in particular – both help to explain the critical organizational capability of market sensing. Second, they are concerned with understanding organization-wide phenomena such as organizational culture and norms. Finally, both encompass relationships and interdependencies between individuals and groups and the co-ordinated use of both tangible and tacit resources. Sinkula (1994) criticizes the fact that there has been relatively little scholarly research on organizational learning in a marketing context and proposes the concept of 'market-based' organizational learning. Writing eight years later, Bell, Whitwell, and Lucas, (2002: 71) contend that a 'number of researchers have emphasized the relevance of organizational learning in several marketing areas' and that 'marketing has a large stake in organizational learning'. Indeed, many researchers view organizational learning as critical to the process of developing market knowledge and, as such, is a driving force of action in, and governance of, market-oriented organizations (ibid.). According to Hurley and Hult (1998: 42) it was Sinkula (1994) and Slater and Narver (1995) who introduced the organizational learning construct to marketing, which represented an important shift in the stream of research on market orientation.

Slater and Narver (1995: 67, 71) contend that market orientation is 'the principle cultural foundation of the learning organization'[17] and that the marketing function has 'a key role to play in the creation of a learning organization'. According to them, learning organizations 'continuously acquire, process, and disseminate throughout the organization knowledge about markets, products, technologies, and business processes', with their knowledge being based 'on experience, experimentation, and information from customers, suppliers, competitors, and other sources' (Slater and Narver, 1995: 71). Day (1994a, 1994b) more or less turns the causality between organizational learning and market orientation around when he suggests that a market-oriented or market-driven approach can emerge only if learning processes are examined and altered in a way that enables firms to 'learn to learn' about markets (cf. also Bell, Whitwell, and Lukas, 2002). In a series of studies, Baker and Sinkula (Baker and Sinkula, 1999a, 1999b, 2005; Sinkula, Baker and Noordewier, 1997) describe learning orientation and market orientation as related but distinct organizational characteristics and examine the link with organizational performance. In their view, market orientation primarily facilitates adaptive learning (see also below) while learning orientation is seen as a mechanism by which generative learning occurs (Bell, Whitwell, and Lucas, 2002). They argue that learning orientation 'can lead an organization astray if a strong market orientation is not present to provide grounding' (Baker and Sinkula, 1999a: 412). One of the major advantages of an enhanced learning orientation is that organizational members 'will not only gather and disseminate information about markets but also constantly examine the quality of their interpretive and storage functions and the validity of the dominant logic that guides the entire process' (Baker and Sinkula, 1999a: 416).

Referring to Fiol and Lyles (1985), Huber (1991), Simon (1969), and Sinkula (1994), Slater and Narver (1995: 63) contend that, at its most basic level, 'organizational learning is the development of new knowledge or insights that have the potential to influence behavior', and – following Senge (1990) – distinguish between 'adaptive learning' (also referred to as single-loop learning by Argyris, 1977) and 'generative learning' (also referred to as double-loop learning by Argyris, 1977). According to Sinkula (1994: 36) – who attempts to characterize the relationship between market information processing and organizational learning – understanding the nature of organizational learning is 'critical to our understanding of how organizations process market information'. In fact, authors in the field of market orientation and organizational learning (see above) have concluded that market-oriented organizations tend to exhibit the behavioural characteristic of seeking to exploit new sources of knowledge and thus conduct generative or double-loop-learning (Chaston, 2004).

Based on Huber's (1991: 90) four organizational learning-related constructs – knowledge acquisition, information distribution, information interpretation, and organizational memory – Sinkula (1994) depicts market information processing – that is, organizational learning in marketing – as a four-stage process that includes information acquisition, information dissemination, shared interpretations, and storage (organizational memory) (cf. also Moorman, 1995; Slater and Narver, 1995). He further proffers that market information processing 'is a function of what the organization has learned in terms of both facts about its relevant markets and its particular way of acquiring, distributing, interpreting, and storing information' (Sinkula, 1994: 37). Research by Moorman (1995: 330) demonstrates that 'information processes may act as "knowledge assets" that can be leveraged to achieve competitive advantage in new products'.

Information acquisition. Information may be acquired from direct experience, the experiences of others, or organizational memory.[18] Indeed, effective managers establish multiple credible internal and external sources to obtain objective information about their enterprise and its surrounding environments (Slater and Narver, 1995: 64). Organizations must continually balance between learning from exploitation – acquiring knowledge from internally focused experience – and exploration – learning from externally focused experience – because too much reliance on the former is unlikely to lead to generative learning, whereas too much reliance on the latter is expensive and may produce too many underdeveloped concepts and ideas (Kyriakopoulos and Moorman, 2004; March, 1991; Slater and Narver, 1995). More will be said on exploitation and exploration in Chapter 4.2.

Information dissemination. Organizational learning is distinguishable from personal learning by information dissemination and accomplishing a shared (organizational) interpretation of the information (Slater and Narver, 1995: 65). Effective dissemination, or sharing, increases information value when each piece of information can be seen in its broader context by all organizational players who might use or be affected by it and who are able to feed back questions, amplifications, or modifications that provide new insights to the sender (Glazer, 1991; Quinn, 1992; Slater and Narver, 1995).

Shared interpretations. The final stage of organizational learning is shared interpretation of the information (Slater and Narver, 1995: 65). For organizational learning to occur in any business unit, there must be a consensus on the meaning of the information and its implications for that business (Day, 1994a; cf. also Slater and Narver, 1995). Besides, to ensure that all information is considered, organizations must provide forums for information exchange and discussion (Slater and Narver, 1995).

Storage (organizational memory). Organizational memory is the fundamental result of organizational learning and organizations use memory as a market information filter (Sinkula, 1994: 41–2; cf. also Walsh and Ungson, 1991). Besides, without organizational memory, learning would have a relatively short half-life because of personnel turnover and the passage of time (Sinkula, 1994; Slater and Narver, 1995). According to Bell, Whitwell, and Lucas (2002: 74), memory is seen as 'a repository for what has been learned in the past' and the term 'organizational knowledge' (cf. Chapter 3.4) has been used to describe a similar function. According to Day (1994a), market sensing relies on organizational memory (stored knowledge and mental models) to facilitate the acquisition and interpretation of incoming market information, as an organization disseminates and utilizes this information to facilitate strategic action, then evaluates that action and updates its organizational memory after observing the outcome (cf. also Kyriakopoulos and Moorman, 2004).

In contrast to, for example, Garvin (1993, 2003), Slater and Narver (1995) do not include behaviour change as an element in the learning process. It is possible, however, that new knowledge confirms what was already suspected or changes managerial perspectives (Menon and Varadarajan, 1992). Consequently, behaviour may not change, but may be pursued more confidently as a result of the new knowledge, or the stage may be set for some future behaviour change to occur (Sinkula, 1994). Whether behavioural change is actually part of the learning process or a separate and distinct activity is less important than recognizing that, in the long term, behaviour change is an essential link between learning and performance improvement (Fiol and Lyles, 1985; cf. also Slater and Narver, 1995). Last, but not least, Hult and Ferrell (1997) apply Garvin's (1993: 80) definition that a learning organization is 'skilled at creating, acquiring, and transferring knowledge, and at modifying its behavior to reflect new knowledge and insights' (cf. Chapter 3.2) to market-driven learning organizations and (global) market knowledge.

Key indicators of market orientation allegedly include the organization-wide gathering of information followed by its interdepartmental dissemination, consideration, and processing, and the organizational use of this information to respond to change (Kohli and Jaworski, 1990). Slater and Narver (1995) suggest that the capacity to react quickly and effectively to outside change depends on a deep understanding of external environments and the free exchange and flow of information to ensure that expertise is available where and when it is required. Hence, they argue, market orientation constitutes a critical input to the idea of the learning organization because it presupposes extensive customer and competitor research, the internal spreading and employment of information to improve performance, and the integration of functions in order to gain knowledge, innovate, and react quickly to market change (Bennet and Gabriel, 1999). This ability gives the market-driven business an advantage in the speed and effectiveness of its response to opportunities and threats, a fact that leads Slater and Narver (1995: 67) to the conclusion that 'a market orientation is inherently a learning orientation'.

As mentioned above, the 'ability of the firm to learn about customers, competitors, and channel members in order to continuously sense and act on events and trends in present and prospective markets' is critically important (Day, 1994a: 43; cf. also Day, 1994b). Indeed, market-driven firms are 'distinguished by an ability to sense events and trends in their markets ahead of their competitors' and this 'anticipatory capability is achieved through open-minded inquiry, synergistic information distribution, mutually informed interpretations, and accessible memories' (Day, 1994a: 44). This market sensing capability 'determines how well the organization is equipped to continuously sense changes in its market and to anticipate the responses to marketing actions' (Day, 1994a: 49).[19]

However, learning is 'more than simply "taking in information"' and the learning process 'must include the ability of managers to ask the right questions at the right time, absorb the answers into their mental model of how the market behaves, share the new understanding with others in the management team, and then act decisively' (Day, 1994b: 9). In fact, effective learning about markets is 'a continuous process that pervades all decision' and learning processes in market-driven firms are distinguished by (Day, 1994b: 10; cf. also Day, 1994a, 1999a):

- open-minded *inquiry*, based on the belief that all decisions are made from the market back;
- widespread *information distribution* that ensures relevant facts are available when needed;
- mutually informed mental models that guide *interpretation* and ensure everyone pays attention to the essence and potential of the information;
- an accessible *memory* of what has been learned so the knowledge can continue to be used.

Further learning then occurs when the outcomes of the action are systematically evaluated, i.e. errors are detected, judgements confirmed or disconfirmed, and information gaps identified. These insights go to augment the organizational memory and trigger further inquiries (Day, 1994b: 11). Furthermore, Day (1994b: 24) reminds us that 'market learning happens throughout the firm whenever employees come in contact with customers, whenever service people solve problems, or whenever sales people listen to distributors' complaints', while at the same time, 'learning also means unlearning obsolete market knowledge'.

Finally, to be 'a powerful foundation for a learning organization and provide the opportunity for generative learning, the scope of market orientation must include all stakeholders and constituencies that (1) possess, or are developing, knowledge that has the potential to contribute to the creation of superior customer value or (2) are threats to competitive advantage' (Slater and Narver, 1995: 68). Therefore '[t]he conception of "market" should be broadened to encompass all sources of relevant knowledge and ideas pertaining to customers and customer value creating capabilities' (ibid.). Consequently, the next section will explore the nature of such a market and marketing knowledge.

4.1.2 Marketing knowledge

'Knowledge – most notably market knowledge, which is directly related to market information about customers, competitors, suppliers, distributors, and so forth, and internal knowledge, such as technology or specialized skills of operation – is a strategically important resource of a firm, and it serves as a basic source of competitive advantage' (Cui, Griffith, and Cavusgil, 2005: 34). According to Hanvanich, Dröge, and Calantone (2003: 125), some marketing scholars have approached marketing knowledge as 'market orientation' (for example, Jaworski and Kohli, 1993; Sinkula, Baker, and Noordewier, 1997; Slater and Narver, 1995). Indeed, as discussed above, Kohli and Jaworski (1990) suggest that firms can gain knowledge or generate market intelligence through the use of market research and they relate this type of research not only to market, sales, pricing, promotion, and customers, but also to environmental scanning. This chapter explores different concepts of marketing knowledge in the relevant literature.

4.1.2.1 *What is marketing knowledge?*

As we have seen in the previous chapters, the emergence of the knowledge society has led to a re-evaluation of the concept of knowledge when applied to management in general and marketing in particular. In fact, as the 'marketing information revolution' (Wierenga and Ophuis, 1997: 275) is producing enormous amounts of data, the need for knowledge-based approaches to marketing becomes obvious. 'Firms embedded in ever-changing, competitive environments need to continuously acquire and utilize timely and relevant information to discover and take advantage of opportunities, and to avoid threats that may arise' and to do so 'they need to acquire knowledge about how to act, for example, how to analyse competitors and customers, how to negotiate, how to achieve competitive advantage, and so on' (Ottesen and Grønhaug, 2004b: 522).

In his work on marketing in an information-intensive environment, Glazer (1991: 2) is surprised that 'despite the wealth of evidence that "information" and information technology are rapidly transforming almost all phases of economic and business activity, relatively little formal attention has been paid to the effects of the transformation on marketing theory and practice'. This book is an attempt to reflect this transformation in marketing theory and practice.

No marketing scholar or practitioner is likely to disagree with Rossiter's (2001: 9) pronouncement to the effect that marketing knowledge 'is absolutely fundamental to our discipline'. In fact, '[k]nowledge is increasingly recognized within marketing management as a critical resource that can be managed to enhance the competitive position and financial performance of a firm' (Darroch and McNaughton, 2003: 572) and, as shown above (4.1.1), acquiring knowledge about customers and competitors and sharing this information between functional areas within a firm are key dimensions of a market orientation. Deshpandé (2001: 1) contends that 'no knowledge is as critical to management, or as elusive, as knowledge about customers, competitors, and markets'. Yet a review of the scholarly treatment of the notion of marketing knowledge makes it clear that this re-evaluation is by no means a straightforward task. Indeed, as the following discussion will make clear, there is no unified view as to the nature and scope of marketing knowledge. As Grønhaug (2002: 364) puts it, marketing knowledge 'comes in many forms'. As will be discussed in greater detail below (Chapter 6.3), (marketing) knowledge plays an essential role in the service-dominant view as proposed by Vargo and Lusch (2004: 2) who define services 'as the application of specialized competences (knowledge and skills) through deeds, processes, and performances for the benefit of another entity of the entity itself'. Indeed, two of their eight foundational premises for their service-centred view are (1) the application of specialized skills and knowledge as the fundamental unit of exchange and (2) knowledge as the fundamental source of competitive advantage (Vargo and Lusch, 2004). Consistent with what has been said about knowledge assets (see Chapter 3.4.3), Glazer (1991) sees marketing knowledge as a strategic asset.

According to Schlegelmilch and Chini (2003: 226), 'it is high time to include marketing into the research agenda for knowledge management and to overcome the paradox that marketing functions are neglected in the discussion on knowledge transfer'. As a matter of fact, '[b]eing among the first to internationalize, marketing functions are key to the development of knowledge transfer processes in a dispersed

MNC context' (ibid.). However, despite the obvious importance of knowledge to the marketing discipline, the marketing literature has struggled for more than ten years to come to an understanding of the nature of marketing knowledge and there does not seem to be a common ground unifying scholars (Kohlbacher, Holden, Glisby, and Numic, 2007). In fact, even though 'marketers have been using knowledge management long before this latter phrase began to be popularized in the management literature' (Chaston, 2004: 22–3) there has to date been 'no clear statement about the forms that marketing knowledge can take, or its content' (Rossiter, 2001: 9). Simonin (1999a: 464) speaks of the 'strategic significance' of marketing know-how, and argues that – in research on international alliances – with only a few exceptions dealing with local market knowledge (Inkpen and Beamish, 1997; Makino and Delios, 1996), 'marketing skills and know-how have yet to receive proper conceptual and empirical attention as a competency source of competitive advantage'. Bjerre and Sharma (2003: 125) note that the 'concept of marketing knowledge is defined differently by researchers' and that it 'may be different from other types of knowledge'.

Kohlbacher and fellow researchers (2007) conclude that '[a]ll in all, it may not be outlandish to suggest that the marketing discipline is tying itself up in semantic knots while it struggles to create consensus on an agreed definition of the term "marketing knowledge", its practical scope and supporting constructs'. While Bjerre and Sharma (2003: 140) posit that the 'important thing is not one specific piece of knowledge, but an entire package that includes knowledge about clients, competitors, local institutions, suppliers etc.', Hanvanich, Dröge, and Calantone (2003: 124–5) observe that both marketing scholars and knowledge management practitioners 'face difficulty in defining what knowledge and marketing knowledge is' and that there is 'no consensus as to how marketing knowledge should be defined and measured'. For Simonin (1999a), for example, marketing knowledge is an organizational resource that reduces the effects of ambiguity and complexity in cross-border interactions. In contrast, Achrol and Kotler (1999: 157) argue that marketing knowledge is a 'primary source of coordinating power' in business networks. From a different point of departure, Menon and Varadarajan (1992) remind us that the marketing discipline can view knowledge according to its focus, scope, process, scale, type, and extent, noting that knowledge constructs can be unidimensional and multidimensional, and note that objective and subjective factors influence perceptions of its task-specific value and relevance, and so forth. Richards, Foster, and Morgan (1998: 48) plough yet a different furrow, stating that knowledge 'is the essence of what a brand represents, how it can achieve competitive advantage and ultimately significant value to a business' and that brands 'are, quintessentially, knowledge'. This seems to be in line with the argumentation of Akutsu and Nonaka (2004) who – using the theory of organizational knowledge creation and an extended notion of brand knowledge – redefine the brand-building method as the brand knowledge-creation process. But probably many marketers would consider this notion of brands as knowledge to be both a hyperbole and a restrictive notion of knowledge in the wider context of marketing (Kohlbacher, Holden, Glisby, and Numic, 2007).

4.1.2.2 Market knowledge

Before exploring the nature of marketing knowledge in greater detail, I will briefly discuss a related concept, namely 'market knowledge' (cf., for example, Eriksson and Chetty, 2003; Li and Calantone, 1998; Marinova, 2004; Sinkula, 1994; Sinkula, Baker, and Noordewier, 1997; Vicari and Cillo, 2006). For many years, researchers and managers have focused their attention on the role of technological knowledge in the innovation process, somehow neglecting the role played by market knowledge (Brown and Eisenhardt, 1995; Verona, 1999; Vicari and Cillo, 2006), a fact that actually also holds true for marketing knowledge. But some researchers have tried to fill this gap and attempt to consider both on a conceptual and an empirical basis the impact that market knowledge might have on innovation (Li and Calantone, 1998; Vicari and Cillo, 2006). Li and Calantone (1998) operationalized 'market knowledge competence', which encompassed customer knowledge process, marketing-R&D interface and competitor knowledge (cf. also Hanvanich, Dröge, and Calantone, 2003; Li and Cavusgil, 1998; Yeniyurt, Cavusgil, and Hult, 2005). For Marinova (2004: 3), 'market knowledge implies knowledge about customers and competitors'. Others also emphasized the importance of learning from the market and market knowledge (cf., for example, Doz, Santos, and Williamson, 2001, 2003; Leonard, 1998, 2000; Santos, Doz, and Williamson, 2004) and according to Doz, Santos, and Williamson (2003: 158), market knowledge includes knowledge on how to serve consumers that behave in a certain way, or what consumers value in a product. However, it is important to note that 'the half-life of usable market knowledge shrinks in the face of compressed life cycles, fragmented markets, and proliferating media and distribution channels', which is why 'it is becoming much harder to stay well-educated' (Day, 1994b: 9). But in order to 'react to market changes quickly, MNC subsidiaries need a sufficient accumulation of previous knowledge and must be able to obtain, process, and apply new market knowledge both quickly and effectively' (Cui, Griffith, and Cavusgil, 2005: 38).

Simonin (1999a) builds his concept of marketing know-how on Hitt and Ireland's (1985) marketing knowledge construct and Johanson and Vahlne's (1977) and Choi and Lee's (1997) concept of market knowledge and proffers that it is related more closely to procedural than to declarative knowledge (see, for example, Kogut and Zander, 1993; Nonaka, 1994). Johanson and Vahlne (1977: 26) define market knowledge as 'information about markets, and operations, in those markets, which is somehow stored and reasonably retrievable – in the mind of individuals, in computer memories, and in written reports'.[20] Choi and Lee's (1997: 43) more elaborate definition runs like this:

> Market knowledge: Knowledge held by consumers as well as firms in the market. Due to the nature of market transactions, knowledge available in the market tends to be highly codified and explicit, but there can be a certain degree of tacit and culture-specific knowledge, such as consumer preferences. Organizations often acquire and utilize market knowledge through intermediaries such as advertising agencies, market research firms, and consulting firms.

Pollard (2006: 6) reminds us that 'it is necessary to distinguish between market knowledge, i.e. details about the market being entered or considered for entry, and marketing knowledge, which includes the ability to process and apply relevant information to deal with markets effectively, whether domestically or in dealing with the challenges of a new foreign market'. Indeed, in contrast to their development of knowledge concerning specific markets (market knowledge), firms require adequate marketing skills (marketing knowledge) in order to exploit fully their market opportunities. Moreover, two levels of market knowledge can be distinguished (Vicari and Cillo, 2006).[21] The first one is the knowledge that a company has about the actors in the market, i.e. customers, trade, and competitors, etc. (Day and Nedungadi, 1994). The second one is the knowledge that customers and trade have and that may be usefully deployed by companies through the enactment of specific tools for customer/trade knowledge capturing (Vicari and Cillo, 2006: 187). This second typology of knowledge resides in the interactions a company enables in the market, using different mechanisms to integrate customer and competitor knowledge into its knowledge base (ibid.). Finally, the knowledge asset developed by a company is the result of the integration of these two typologies of knowledge, with the second type – customer and competitor knowledge – being more difficult to generate because of its tacit nature (Vicari and Cillo, 2006: 188). But it comprises also the kind of knowledge that might represent a real source of competitive advantage because it enables the firm to satisfy expressed and latent needs and to foster its innovative activity while leveraging a high-potential knowledge that is its market network's knowledge (Jayachandran, Hewett, and Kaufmann, 2004; cf. also Vicari and Cillo, 2006).

As will be explained in Chapter 4.2.1 and 4.2.2, my notion of marketing knowledge subsumes market knowledge and integrates both levels of it.

4.1.2.3 Academic marketing knowledge and practical marketing knowledge

A review of the relevant literature shows that it is crucial to distinguish between academic and practical marketing knowledge, that is, marketing knowledge used by practitioners (cf., for example, Cavusgil, 1998; Grønhaug, 2002; Hackley, 1999; McIntyre and Sutherland, 2002; Midgley, 2002; Ottesen and Grønhaug, 2004b; Rossiter, 2001, 2002; Wierenga, 2002; Wierenga and van Bruggen, 2000). Indeed, a large dialogue about the development, dissemination and utilization of academic marketing has been going on since the 1980s (cf., for example, AMA Task Force on the Development of Marketing Thought, 1988; Bloom, 1987; Holbrook, 1995; Hubbard, Brodie, and Armstrong, 1992; Leone and Schultz, 1980; McIntyre and Sutherland, 2002; Myers, Massy, and Greyser, 1980; Rossiter, 2001; Varadarajan and Menon, 1993). Especially the American Marketing Association's Task Force on the Development of Marketing Thought (1988) has initiated and strongly contributed to the debate on the relevance of academic marketing knowledge to practitioners, the lack of accumulation of it, and problems of its transfer (cf. also Bloom, 1987; McIntyre and Sutherland, 2002; Varadarajan and Menon, 1993). However, this discussion lost much of its vigour in the 1990s until Rossiter's (2001) article 'What is marketing knowledge?' rekindled it and led to a special issue on 'marketing knowledge' in the journal *Marketing Theory* in 2002. In his article, Rossiter (2001: 9) complains that most works on marketing knowledge fail to provide a definition or explanations of what it consists of. Indeed, the AMA's Taskforce on the Development of Marketing Thought (1988: 4) concluded that there has been 'too little effort directed to systematic development of marketing knowledge', for which Rossiter (2001: 20) sees 'avoidance of the logically prior question of what exactly marketing knowledge is' as the major reason. He therefore deems it necessary to identify the forms that marketing knowledge can take, and contends that marketing knowledge 'is what marketing academics and consultants teach and marketing managers draw upon in formulating marketing plans' (Rossiter, 2001: 9). He further proposes four forms of marketing knowledge – marketing concepts, structural frameworks, strategic principles, and research principles – to which he added a fifth one – empirical generalizations – in another article one year later (Rossiter, 2002).

A detailed discussion of the five forms would go beyond the scope of this book, but a brief explanation seems to be in order. '*Marketing concepts* are the building blocks of marketing knowledge and are needed to understand the other three forms, since these contain concepts. *Structural frameworks* are models that are non-causal – in everyday terminology, they are "useful checklists". *Strategic principles* are hypothesized causal models that relate one concept to another in a functional "if, do" form. *Research principles* are hypothesized causal models pertaining specifically to the appropriate use of particular research techniques' (Rossiter, 2001: 13). 'Empirical generalizations are associations (or correlations) of marketing concepts, and thus differ from the independence of marketing concepts in themselves and the merely nominal relationship between marketing concepts in structural frameworks' (Rossiter, 2002: 372; cf. also Uncles, 2002). Rossiter's (2001, 2002) framework has been deemed too restrictive in different ways by different commentators (Brodie, 2002; Midgley, 2002; Uncles, 2002; Wierenga, 2002).[22] These comments have led to the inclusion of the fifth form of marketing knowledge into Rossiter's framework. In a recent working paper, Rossiter (2005) evolved to the second stage of his marketing knowledge framework. Written as a sequel to his first two articles on marketing knowledge (2001, 2002), Rossiter aims 'to explain and evaluate the types of evidence that it is possible to bring to bear on deciding acceptable content for the forms of marketing knowledge' and proposes four main types of evidence: expert opinion, experience-analogizing, empirical evidence from experiments and surveys, and logical reasoning (Rossiter, 2005: 2). The conclusion is that 'expert opinion has to be the largest-contributing type of evidence for marketing knowledge', which 'is true because of the paucity of evidence, of any type, for most areas of marketing knowledge other than what has been proposed by various experts' (Rossiter, 2005: 12, removed emphasis). Finally, he reminds us that 'all marketing knowledge, just like all other knowledge, is provisional, except the knowledge derived by logic (which is tautologically true by definition)' (ibid., removed emphasis).

Therefore, all that we can hope for is 'best bet' concepts, frameworks, and principles that are true with reasonable probability (Rossiter, 2005). But given the status of his working paper, it seems not outlandish to suggest that Rossiter's framework of marketing knowledge is still a provisional, best bet framework, too.

4.1.2.4 Tacitness of marketing knowledge

One of the shortcomings of Rossiter's (2001, 2002) – as well as other scholars' – concept of marketing knowledge is that he explicitly excludes tacit knowledge. For his purpose, 'marketing knowledge is declarative ("know what") and ... exists independently of, and should be distinguished from, marketing skills or procedural knowledge ("know-how")' (Rossiter, 2001: 10). He further states that marketing knowledge 'must exist independently of practitioners' ability to use it, so that marketing knowledge can be documented and passed on to others' and therefore excludes tacit knowledge 'because of its incapacity to be codified and taught' (ibid.). However, as has also been shown in Chapter 3, '[o]ur explicit knowledge is but the small communicable cap of the iceberg of preconscious collective human knowledge, the vast bulk of which is tacit, unseen, and embedded in our social identity and practice' (Spender, 1996b: 54). Bertels and Savage (1999: 211), using the same popular metaphor, put it like this: 'Our ability to track down explicit knowledge in databases, guidelines or organizational charts is only the tip of the iceberg. An organization's real knowledge is often embodied in the experience, skills, knowledge and capabilities of individuals and groups. Beliefs and metaphors shape it.'

Indeed, the importance and relevance of tacit knowledge in general have already been discussed in Chapter 3. I therefore argue that it is very problematic to exclude tacit knowledge from the definition of marketing knowledge and that it might actually be the essential part of it. As we have seen from Chapter 3, there are mechanisms and ways of transferring – or better, re-creating – tacit knowledge, and managing these processes can be decisive for the competitive advantage for firms. In fact, Midgley (2002) confirms that explicit, that is, codified, knowledge is unlikely to aid the firm in competition and McIntyre and Sutherland (2002: 411) proffer that to leave out tacit knowledge as implementation knowledge 'is to miss a critical component of marketing success'. For Vicari and Cillo (2006: 196), even though 'both codified/explicit and complex/tacit knowledge from the market are relevant', 'marketing researchers have been emphasizing the explicit dimension of market knowledge'.

According to Simonin (1999a: 469), 'marketing know-how is generally characterized by a high degree of tacitness' due to its socially complex nature and it is 'rather difficult to think of an easily-codifiable advertising savoir-faire, explicit success formulas for product launches, or clear, replicable blueprints for international market expansions'. Therefore, 'except for the few instances where marketing know-how can be unequivocally codified ... learning from experience and learning by doing in the presence of knowledgeable partners become an essential condition for circumventing ambiguity and favoring knowledge transfer' (Simonin, 1999a: 483; cf. also Cavusgil, Calantone, and Zhao, 2003). While '[l]arge amounts of marketing knowledge are codified in popular texts and constitute a public discourse', 'much of the knowledge underpinning practical marketing expertise may be tacit, implicit in the day-to-day problem solving of strategic marketing practitioners' and, as such, 'difficult to elicit from experts or to codify in public symbols' (Hackley, 1999: 722).[23] This implies that high-level expertise in marketing involves cognitive performance which goes beyond marketing's codified body of knowledge (ibid.). Indeed, Kohlbacher and fellow researchers (2007) have shown that notions of marketing knowledge habitually focus too strongly on explicit knowledge, even though for international (cross-cultural) marketing it is essential that tacit knowledge is built into constructs of marketing knowledge (cf. also Vicari and Cillo, 2006). Bjerre and Sharma (2003: 125) argue that '[m]arketing knowledge is frequently more experiential than, for example, technical knowledge', and, as such, it is 'opaque and difficult to document' and 'also located with the people and teams positioned at the boundary line between the buying and selling firm'. 'Experiential knowledge' is a concept from Penrose (1995: 53–4) – who distinguishes two kinds of knowledge, namely 'objective knowledge' and 'experience' – and means knowledge that firms accumulate by being active in the market (cf. also Hadley and Wilson, 2003). Like tacit knowledge, it is accumulated based on the premise of learning by doing (Bjerre and Sharma, 2003). Or, as Grant (1996b: 111) has put it, '[t]acit knowledge is revealed through its application'. Johanson and Vahlne (1977: 28) also build their concept of market knowledge on Penrose's two types of knowledge and argue that 'experiential knowledge is the critical kind of knowledge' and that it is especially important for less structured and well defined activities, such as managerial work and marketing. Furthermore, they proffer that market-specific knowledge 'can be gained mainly through experience in the market, whereas knowledge of the operation can often be transferred from one country to another country'.

According to Bjerre and Sharma (2003: 123), 'knowledge is market specific and difficult to codify' and therefore its international transfer is hardly feasible (cf. also Vicari and Cillo, 2006). Besides, '[d]ifferent pieces of marketing knowledge may be located in different people or departments' and '[n]o single person can fully comprehend the entire package of marketing knowledge', which makes the intra-unit transfer of marketing knowledge difficult (Bjerre and Sharma, 2003: 127). The fact that basically in any company, 'critically important knowledge resides in the workplace – on the factory floor, within sales and service organizations that deal directly with customers, at the "bench" in the R&D lab', in short at the 'front lines' of the company (Yasumuro and Westney, 2001: 178), underscores the importance of tacit knowledge and its strategic creation and management (Ichijo, 2006a). As mentioned above, this need to unlock the potential of globally dispersed knowledge has been called 'the metanational imperative' (Doz et al., 2001) and the term 'front-line management' has been used to describe a form of management where 'the workplace is recognized and valued as the center of knowledge creation and in which knowledge-creation resources ... and processes ... are concentrated at the front line of the company' (Yasumuro and Westney, 2001: 178). This type of knowledge, experienced, collected, and generated at the front lines of the company, is also termed 'local knowledge', knowledge that is deeply contextual, practical, and derived from lived experience (Yanow, 2004). Yanow (2004: 12, removed emphasis) defines local knowledge as 'the very mundane, yet expert understanding of and practical reasoning about local conditions derived from lived experience'. In Japan these front lines are frequently referred to as *gemba*, which can be loosely translated as 'the actual spot or place' and according to Womack and Jones (2005b: 19) it is 'the Japanese word for the place in the office or factory where the real work is done'. The *ba* in *gemba* is of course the same Japanese word *ba* introduced in Chapter 3. *Gemba* has become famous in relation to Toyota's principle of *genchi genbutsu* – going to the place to see the actual situation for understanding (Liker, 2004: 224). Obviously, *gemba* is also where knowledge acquisition and creation seems to be the most fruitful and the most important.

This valuable knowledge in the marketplace is 'unique and mostly context-specific [and] often difficult to obtain' (Schlegelmilch and Penz, 2002: 7) and 'the most influential knowledge is likely to be tacit' (Day, 1994b: 10). Nevertheless, this is precisely the kind of knowledge which, if discovered and exploited, can be harnessed to secure competitive advantage (Kohlbacher, Holden, Glisby, and Numic, 2007). In fact, according to Yanow (2004), workers at the organizational periphery possess local knowledge which is organizationally relevant. But because it is possessed by people who are located at a hierarchical and geographic remove from the centre of the organization it is often neglected or even deemed to be inferior (Tsoukas and Mylonopoulos, 2004; Yanow, 2004). Finally it is important to note that '[l]ocal market knowledge ... won't benefit the rest of the company unless it is shared so that other parts can consider its value to them' with '[m]arket knowledge [being] not fully captured in a usable form until the lessons and insights are transferred beyond those who gained the experience' (Day, 1994b: 17, 23; for local market knowledge in alliances see Inkpen and Beamish, 1997; Makino and Delios, 1996). However, individuals are the primary repositories of tacit knowledge, which makes it difficult to unravel and to communicate between sections (Bennet and Gabriel, 1999).

4.1.3 Areas of marketing knowledge creation and application

As mentioned above (see Chapter 3.6), scholars and practitioners around the globe have identified the capability of MNCs to create, combine, and efficiently transfer knowledge from different locations worldwide as an increasingly important determinant of competitive advantage, corporate success and survival. But even though 'marketing functions lend themselves particularly well for an investigation of knowledge transfer within MNCs', 'there is a dearth of research on knowledge transfer in the field of marketing' (Schlegelmilch and Chini, 2003: 220–1). Indeed, hardly any research into the in-house management of marketing knowledge has been completed, in sharp contrast to knowledge management research in other disciplines (Bennet and Gabriel, 1999). As mentioned above (see 4.1.1) marketing academics have concentrated on market orientation, especially with respect to linkages between market orientation and organizational learning (Bennet, 1998; Bennet and Gabriel, 1999) but accounts of marketing from a knowledge management or knowledge-based perspective still seem to be rare.

Apart from Chaston's (2004) knowledge-based marketing and Day's (1994a, 1999a) market-driven organizations and their capabilities, one of the exceptions of a knowledge-based view of marketing are the concept of and literature on 'market knowledge competencies' (for example, Li and Calantone, 1998; Li and Cavusgil, 1998; Yeniyurt, Cavusgil, and Hult, 2005). Yeniyurt, Cavusgil, and Hult (2005) propose a global market advantage framework and explore the role of global market knowledge competencies within it. They build on the resource-based view of the firm (Wernerfelt, 1984, 1995) (see also Chapters 3.3 and 4.1.1) and argue that the one 'specifically pertaining to the application of a firm's idiosyncratic abilities in the attainment of a sustained competitive advantage (Barney, 1991) in a global marketplace provides a strong theoretical foundation for the exploration of global market knowledge competencies and their relative effect on firm performance' (Yeniyurt, Cavusgil, and Hult, 2005: 3). According to their framework, the knowledge management competencies consist of global customer, competitor and supplier knowledge development, inter-functional co-ordination, and value chain co-ordination. As a result, global organizations should possess the capability of acquiring, interpreting, and integrating intelligence (Huber, 1991; Sinkula, 1994) regarding the global trends in customer preferences, the global competitive environment, and global suppliers (Yeniyurt, Cavusgil, and Hult, 2005). Additionally, inter-functional co-ordination (for example, Hult and Ferrell, 1997; Narver and Slater, 1990) and dissemination of all of the above knowledge across various functions of the business unit (Kohli and Jaworski, 1990; Narver and Slater, 1990), as well as global value-chain coordination are critical competencies (Yeniyurt, Cavusgil, and Hult, 2005). Finally, the global market knowledge capabilities 'enable the firm to create a global market advantage, when compared with the competition' (Yeniyurt, Cavusgil, and Hult, 2005: 11).

Hanvanich, Dröge, and Calantone (2003: 124) argue that while marketing scholars have been interested in the topic of marketing knowledge, 'they have focused mainly on how firms acquire, disseminate, and store knowledge', with related research areas being market orientation and organizational learning (see Chapter 4.1.1). Taking a new approach to reconceptualizing marketing knowledge and innovation, Hanvanich, Dröge, and Calantone (2003: 130) proffer that 'marketing knowledge resides in three key marketing processes: product development management (PDM), customer relationship management (CRM), and supply chain management (SCM)'. This notion is based on Srivastava, Shervani, and Fahey's (1999) framework that redefines marketing as a phenomenon embedded in the three core marketing processes of PDM, SCM, and CRM. In addition, Hanvanich, Dröge, and Calantone's (2003) findings also support Bennet and Gabriel's (1999) contention that marketing requires knowledge of customers and their preferences, competitors, products, distribution channels, service providers, laws and regulations, and general management practices. They further argue that 'marketing knowledge is the extent of understanding of these three marketing processes, an extent which can be measured by evaluating awareness of factors, control of factors, and application of knowledge in new markets (each successively requiring more extensive knowledge of PDM, SCM and CRM processes)' (Hanvanich, Dröge, and Calantone, 2003: 130–1). Last, but not least, the main result of their empirical study shows that marketing knowledge was different from, but positively related to, marketing innovation. This positive relationship supports Glazer's (1991) contention that marketing knowledge is a strategic asset since it demonstrates that marketing knowledge is associated positively with the ability to achieve radically superior products, the potential to uncover new demands, and the capability to build competencies through collaboration with other firms (Hanvanich, Dröge, and Calantone, 2003: 131).

I will subsequently discuss each of the three core processes of PDM, SCM, and CRM from a marketing knowledge perspective and also add the process of market research. As will be shown, I view customer knowledge management (CKM) as one – from a knowledge-based perspective, essential – process within CRM. As a matter of interest, all of these processes could also be mapped into Porter's (1980, 1985) value chain. Therefore, in a sense, the following sections look at knowledge-based marketing and marketing knowledge issues along the value chain. In fact, Hult and Ferrell (1997: 155) argue that firms tap 'into the cumulative knowledge of its entire value chain to be market-oriented'.

I propose that SCM, market research, CRM, and product development are interdependent and interwoven processes. They mutually benefit from each other's knowledge and should be managed in an integrated and comprehensive way. This is particularly true for market research, CKM (see 4.1.3.3), and PDM. It is only for the sake of structure and clarity that I treat them separately in different sections in this book.

4.1.3.1 *Supply chain management (SCM)*

SCM might actually be the least obvious process to analyse from a marketing knowledge perspective. However, suppliers may be able to generate and provide valuable insights and knowledge about competitors, customers, and customers' customers, and they can play an important role in product development processes and help to cut costs and provide superior value propositions to customers. Desouza, Awazu, and Jasimuddin (2005: 16) assert that 'organizations must use their suppliers as an avenue to interact with other members of the value chain and competitors'. Indeed, suppliers must be 'prepared to develop team-based mechanisms for continuously exchanging information about needs, problems, and emerging requirements and then taking action', because in a successful collaborative relationship, joint problem solving displaces negotiations (Day, 1994a: 45). Suppliers must also be prepared to participate in the customer's development processes, even before the product specifications are established (ibid.). That is why the channel bonding capability has many features in common with the customer-linking capability, and hence the same skills, mechanisms, and processes might be readily transferred between those related domains (Day, 1994a: 44n).

Ahmadjian (2004: 227) contends that '[k]nowledge creation occurs not only within firms, but also through relationships between firms'. In fact, customer–supplier partnerships (Konsynski and McFarlan, 1990) as well as strong supplier networks have frequently been put forward in this context (cf., for example, Ahmadjian and Lincoln, 2001; Chakravarty, 2000; Chaston, 2004; Cusumano and Takeishi, 1991; Dyer, 1996a, 1996b; Dyer and Hatch, 2004, 2006; Dyer and Nobeoka, 2000; Dyer and Ouchi, 1993; Kotabe, Martin, and Domoto, 2003; Liker and Choi, 2004; Liker and Yu, 2000; Lincoln, Ahmadjian, and Mason, 1998). Supplier networks are particularly strong in Japanese companies, especially those in the automotive and the electronics sectors. These networks or strong relationships between firms in Japan have frequently been termed and analysed as so-called *keiretsu* (conglomerates, called *chaebol* in Korea, cf., for example, Porter and Sakakibara, 2004; Thomas, 2002), described as 'the webs of relations that envelop many Japanese companies' (Lincoln, Gerlach, and Ahmadjian, 1996: 67) or as 'clusters of interlinked Japanese firms and the specific ties that bind them' and their 'long-term, personal and reciprocal character' (Lincoln, Gerlach, and Takahashi, 1992: 561).[24] According to Dyer and Ouchi (1993: 51), 'evidence from an increasing number of industries and sources suggests that much of the Japanese success can be attributed to Japanese-style business partnerships'. Indeed, 'the Japanese style of supply chain management is now worldwide "best practice"' (Desouza, Awazu, and Jasimuddin, 2005: 19). Furthermore, 'the openness and richness of networks are believed to foster a fertile environment for the creation of entirely new knowledge' (Lincoln, Ahmadjian, and Mason, 1998: 241). But it is not necessarily only big firms that successfully manage and share knowledge in the supply chain. Glisby and Holden (2005), for example, present the case of a Danish small specialist manufacturer that applied knowledge management concepts to the supply chain and thus managed to co-create the market with their Japanese business partners through a synergistic process of knowledge sharing (cf. also Kohlbacher, Holden, Glisby, and Numic, 2007).

In a similar vein, Chaston (2004: 15) argues that as 'organizations come to appreciate the value of acquiring new knowledge as the basis for gaining competitive advantage, new intra- and interorganisational structures are beginning to emerge to provide mechanisms for delivering new, more entrepreneurial business strategies'. The spider's web appearance of such forms of collaboration has resulted in the emergence of terminology of 'knowledge networks' or 'learning networks' to describe these new organizational forms (Chaston, 1999). But these networks can also take the form of so-called 'cascade knowledge networks' (Chaston, 2004). Within these networks, the OEM accepts the role of guiding and resourcing the learning process within their market system (Chaston, 2004). Chaston (2004: 21) concludes that 'in the twenty-first century, it can confidently be predicted that knowledge networks of various forms will become an increasingly dominant operational structure through which to ensure the effective management of entrepreneurial activities in both private and public sector organisations'. More will be said on business networks from a knowledge-based marketing perspective in Chapter 4.2.2.2.

4.1.3.2 Market research

Specifically in the 1980s and the beginning of the 1990s, there was a 'growing recognition within both the academic and the business communities of the importance of the study of the knowledge utilization process in marketing and other administrative disciplines', and in the 'drive by corporations to become more competitive and more market oriented, utilizing market intelligence and marketing-research-generated information has gained center-stage status' (Menon and Varadarajan, 1992: 53). Indeed, as discussed above, Kohli and Jaworski (1990) suggest that firms can gain knowledge or generate market intelligence through the use of market research and they relate this type of research not only to market, sales, pricing, promotion, and customers, but also to environmental scanning. They therefore characterize and define 'organizationwide *generation* of market intelligence ... *dissemination* of the intelligence across departments, and organization-wide *responsiveness* to it' as the critical elements of market orientation (Kohli and Jaworski, 1990: 6, original emphasis). 'Importing knowledge from the market is clearly an essential activity in the design of a range of product lines – including some that meet current demand and others that anticipate future customer needs' (Leonard, 1998: 211). Grundei (2000: 342) sees market research as a knowledge management function, arguing that marketing research 'has a key position within ... market related knowledge management'.

In the marketing literature, the terms 'knowledge use' and 'knowledge utilization' usually mean research utilization or research knowledge utilization (Menon and Varadarajan, 1992). Menon and Varadarajan (1992), based on an extensive literature review, propose a framework for circumscribing the concept of marketing knowledge utilization in firms and present a conceptual model and research propositions delineating the relationship between key organizational and informational factors and marketing knowledge utilization in firms. They finally conclude that 'though the *characteristics of knowledge* are important determinants of its utilization, the *characteristics of the firm* (i.e., the knowledge user) are just as important', if not more important (Menon and Varadarajan, 1992: 68, original emphasis). However, a detailed discussion of the use and utilization of marketing research and marketing research knowledge in firms is beyond the scope of this book.

Traditionally, market research was used to shed more light on what the customer knew and thought about the product, and how this differed from what the company had to offer the customer, resulting in enormous CRM databases (see Gibbert, Leibold, and Probst, 2002; Wikström, 1996b; Woodruff, 1997). Indeed, in marketing, much knowledge about consumer decision-making is based on information gathered through verbal protocols (telephone interviews, group meetings, questionnaires) that rely on self-reflection and self-awareness. But 'these methods are largely confined to seeing only what is on the tip of the iceberg' because so much of our knowledge is 'unconscious or tacit that we can never be fully aware of all that we know' and since 'most knowledge is hidden, surfacing it presents a major challenge' (Zaltman, 2003: 40–1; cf. also Furukawa, 1999a). In fact, 'customers, like employees, are often not able to make knowledge, i.e. their experiences with the company's products, their skills, and reflections explicit, and thereby easily transferable and shareable' (Gibbert, Leibold, and Probst, 2002: 461). For example, salespeople's knowledge about customers is often tacit in that it is personal, anecdotal, and situationally prescribed. Such knowledge, according to Clippinger (1995: 28), is 'typically neither created nor shared through traditional channels, rather it emerges and evolves from the bottom up in a somewhat helter-skelter pattern'. Edvinsson and Sullivan (1996) similarly point out that much customer relationship knowledge is tacit and transferred via conversation and on-the-job training (and also that it is not protected by intellectual property law so that once transferred there are few means for the original owner to reassert ownership) (cf. also Bennet and Gabriel, 1999).

Therefore, by any measure, 'traditional market research provides only a small part of available knowledge about consumers' (Zaltman, 2003: 240; cf. also Furukawa, 1999a), which is why Kagan (2002) is critical of questionnaire data in ways that apply to marketing, including their inability to reveal tacit or unconscious knowledge. Garvin (2003: 145) even concludes that market research 'is often incomplete and misleading because consumers lack a firm basis for describing their preferences or predicting future behavior'. In a similar vein, Vandermerwe (2004: 28) contends that '[m]any organizations rely heavily on market research, but market research is little help in creating a future that customers have yet to imagine' and Leonard (1998: 189) concurs by stating that '[a]t the same time that "listening to the customer" has become an important management mantra in many companies, the mechanisms for interacting with the market, and especially for obtaining guidance for new-product development, have come under fire'. 'Traditional tools for importing market knowledge are valuable but limited in situations in which technological potential outstrips user understanding' (Leonard, 1998: 211). The same seems to be true for research in marketing, which is 'often limited to collecting standardized data on consumers, competitors and others without getting beyond statistical or verbal description' (Gummesson, 2005: 318). Indeed, 'marketing knowledge can *only in special respects* be built on surveys and detailed studies of the meaning of single concepts – such as commitment and trust – and statistically significant cause-and-effect links' (Gummesson, 2003a: 483, added emphasis). Finally, current methods used to segment markets, build brands, and understand customers have also been deemed inappropriate to create products that customers will consistently value (Christensen, Cook, and Hall, 2005).

Zaltman (2003: 263) highlights the 'essential role of questions in developing marketing knowledge' and contends that 'a marketing manager's questioning strategy shapes his ultimate learning about consumers'. Here, metaphors can be useful for identifying and learning about customer needs. Zaltman (1996, 1997, 2003) building on his work on market research and market information (Barabba and Zaltman, 1991) and an understanding of how the unconscious mind works, has developed a metaphor-elicitation technique[25] for exploring people's largely unconscious feelings about a product or experience. Metaphors do not exist as words in memory, but as networks of abstract understandings that constitute part of our mental imagery (Mitchell, 1994; Zaltman, 2003). According to Zaltman (2003: 92, removed emphasis), 'metaphors are the primary means by which companies and consumers engage one another's attention and imagination' and the 'significance of metaphors for marketing managers comes from their centrality to consumers' imagination'. Indeed, 'as companies attempt to offer complete solutions instead of stand-alone products and services, they need to obtain a better understanding of the myriad, subtle and often unarticulated needs of customers' (Santos, Doz, and Williamson, 2004: 33). Particularly important in the metaphor-elicitation process are the so-called 'core' metaphors, which are deep, tacit, and even unconscious, but are useful in generating ideas for new products, or the positioning of existing ones (Zaltman, 2003). Indeed, '[b]y evoking and analyzing metaphors from consumers, marketers can draw back the curtains on consumers' tacit knowledge, encourage consumers to look in, and then share what they see so that managers can create enduring value for customers in response to the insights revealed' (Zaltman, 2003: 41). Understanding core metaphors is also helpful in strengthening a company's brand and image (Leonard, 2006; Zaltman, 2003).

Finally, '[u]nderstanding market needs is one of the most critical knowledge management tasks for developers of new products and services' (Leonard, 2000: 223). According to Leonard (2006: 146), the 'greatest challenge in product and service innovation is to match what customers will buy to what the organization can produce' and usually the knowledge requisite to accomplish that task resides in two different contexts: the users' and the organization's developers. However, it is far easier to deliver information and data about market segments than knowledge about what customers really need, what they are thinking, or what unconscious motives are driving their behaviour. This is why traditional – mostly large sample, survey-based – market research faces major limitations and recently some non-traditional research techniques that attempt to break these barriers and provide real knowledge – not just information or data – have received increased attention. Basically, all these types of market research are better tools for generating new product and service ideas than they are at testing those ideas (Leonard, 2006; for an overview of market research techniques see Leonard, 1998).

Obviously, market research and CKM are – or at least should be – closely related. In a similar vein, customer knowledge and its management are not only essential for understanding customers and their needs to successfully advertise and market products but also as early as in the product development process. Indeed, as marketing organizations serve as corporate links between customers and the organization's manufacturing and R&D operations (Riesenberger, 1998), the integration of, and knowledge exchange between, R&D and marketing have also been treated as important issues (for example, Griffin and Hauser, 1996; Song and Parry, 1993). In this context, the way of capturing customer needs and translating them into a product concept has been termed 'empathic design' (Leonard, 1998; Leonard and Rayport, 1997; Leonard and Swap, 2005b, see below).

4.1.3.3 Customer relationship management (CRM)

The importance of customer knowledge has already been emphasized in Chapter 4.1.1 on market orientation and the terms 'market-oriented', 'market-driven', and 'customer-focused' have been used interchangeably (see also above, 4.1.1).[26] Especially the continuous need to learn from and about customers and competitors and to exploit such knowledge to stay ahead has frequently been stressed and discussed (cf., for example, Chaston, 2004; Davenport, Harris, and Kohli, 2001; Davenport and Klahr, 1998; Gulati and Oldroyd, 2005; Li and Cavusgil, 1998). The significance of this kind of learning about and from customers is depicted vividly in a case study of Seven Eleven Japan in an article on market knowledge by Nonaka and fellow researchers (1998).[27] Ogawa (2000) proposes the concept of 'demand chain management' (DCM) – with chain referring to chain operation – which is a management model of customer correspondence and attendance and makes active use of knowledge from outlets and customer knowledge and shares this knowledge throughout the organization. Obviously, market sensing, customer linking, and channel bonding capabilities (Day, 1994a) play an important role in this context (cf. also 4.1.1).

Indeed, 'customer focus', 'customer knowledge co-creation', 'customer interaction', and 'customer intimacy' are crucial keywords in this context (cf., for example, Griffin and Hauser, 1993; Gruner and Homburg, 2000; Gulati and Oldroyd, 2005; Homburg, Workman Jr., and Jensen, 2000; Katahira, Furukawa, and Abe, 2003; Lawer, 2005; Prahalad and Ramaswamy, 2004a; Thomke and von Hippel, 2002; Treacy and Wiersema, 1993; Vandermerwe, 2000, 2004).[28] Customer knowledge co-creation – including the lead user concept – will be discussed in Chapter 4.2.3.

In their recent article on customer focus, Gulati and Oldroyd (2005: 94) argue that companies can become customer focused 'only if they learn everything there is to learn about their customers at the most granular level, creating a comprehensive picture of each customer's needs – past, present, and future'. However, '[t]o be of greatest use, customer information must move beyond the market research, sales, and marketing functions and "permeate every corporate function" – the R&D scientists and engineers, the manufacturing people, and the field-service specialists' and 'regular cross-functional meetings to discuss customer needs and to analyse feedback from buying influences are very important' (Shapiro, 1988: 120). Homburg, Workman Jr., and Jensen, (2000: 467) define a customer-focused organizational structure as 'an organizational structure that uses groups of customers related by industry, application, usage situation, or some other nongeographic similarity as the primary basis for structuring the organization'. They point to the fact that there is a difference between the idea of a customer-focused organizational structure and the idea of a market-oriented organization, as market orientation is primarily about cultural and behavioural aspects (cf. above, 4.1.1) rather than structure. They view a customer-focused organizational structure as an antecedent to and as a facilitator of market orientation. Factors other than market orientation encouraging the development of knowledge management include advances in information technology (allowing companies to accumulate vast amounts of information on customer and market characteristics) and the general broadening of the typical business executive's role to incorporate a wider variety of tasks, hence stimulating his or her demand for knowledge. In the marketing sphere the latter consideration might be especially relevant vis-à-vis relationship marketing, integrated marketing communications, customer support and liaison, database management, and new product development (Bennet and Gabriel, 1999). Wikström and Norman (1994: 64) argue that because marketing is no longer 'a clearly delineated function at the end of the production chain' and that since nowadays 'there are many functions and people who influence the customer relationship', it is not logical to have marketing handled solely by a specialist department. Thus, knowledge about customers needs to be shared throughout the organization (Bennet and Gabriel, 1999). Indeed, 'knowledge on customers and their preferences must be located or solutions for a particular kind of customer problem need to be identified' (Schlegelmilch and Penz, 2002: 12). For the latter task, CRM and data mining tools for decision support have proven useful (Shaw, Subramaniam, Tan, and Welge,

2001; Wierenga and Ophuis, 1997; for recent developments in the field of automated decision-making see Davenport and Harris, 2005) and effective CRM is 'critically dependent upon having accurate and up-to-date knowledge about customers' (Chaston, 2004: 225). CRM 'allows the company to discover who its customers are, how they behave, and what they need or want' and it also 'enables the company to respond appropriately, coherently, and quickly to different customer opportunities' (Kotler, Jain, and Maesincee, 2002: 28). But even though tools and technology are important, they are not enough (cf., for example, Davenport, Harris, and Kohli, 2001; Day, 2003; Gulati and Oldroyd, 2005). As Dixon (2000: 5) puts it: 'Technology has to be married with face-to-face interaction to create the most effective systems; one does not replace the other, although clearly one can greatly enhance the other.'

In fact, although CRM has received much scholarly and management attention (for example, Berry and Linoff, 1999; Curry and Curry, 2000; Day, 2003; Desouza and Awazu, 2005b; Fournier, Dobscha, and Mick, 1998; Parvatiyar and Sheth, 2000; Peppers and Rogers, 1997, 1999; Peppers, Rogers, and Dorf, 1999; Pine II, Peppers, and Rogers, 1995; Shaw, Subramaniam, Tan, and Welge, 2001; Webster, 2002), it frequently does not go beyond the surface and remains restricted to collecting and managing mere data and information, but not knowledge – especially tacit knowledge – despite the importance identified in Chapters 3 and 4.1.2.4 (cf. also Zaltman, 2003). Indeed, Gouillart and Sturdivant (1994: 117) complain that 'most managers do not understand the distinction between information and knowledge' and that even if they 'include information from all points on the distribution channel, most general market data do not show a manager how each customer relates to the next or how customers view competing products and services'. Besides, CRM has been traditionally popular as a means to tie customers to the company through various loyalty schemes, but it left perhaps the greatest source of value under-leveraged: the knowledge residing in customers (Gibbert, Leibold, and Probst, 2002: 464).

A relatively new approach that tries to overcome the shortcomings of CRM (cf., for example, Fournier, Dobscha, and Mick, 1998; Rigby and Ledingham, 2004; Rigby, Reichheld, and Schefter, 2002; Seybold, 2001) is 'customer knowledge management' (CKM) (for example, Davenport, Harris, and Kohli, 2001; Desouza and Awazu, 2004, 2005a, 2005b; Gibbert, Leibold, and Probst, 2002; Leibold, Probst, and Gibbert, 2002; Sawhney and Prandelli, 2000a; Murillo-García and Annabi, 2002; Wayland and Cole, 1997). According to Gibbert, Leibold, and Probst

(2002: 461) CKM differs from CRM and knowledge management in general, as customer knowledge managers 'require a different mindset along a number of key variables' (see Figure 4.2). 'Customer knowledge managers, first and foremost focus on knowledge *from* the customer (i.e. knowledge residing in customers), rather than focusing on knowledge *about* the customer, as characteristic of customer relationship management' (Gibbert, Leibold, and Probst, 2002: 461, original emphasis). Indeed, customer-driven companies need to harness their capabilities to manage the knowledge of those who buy their products (Baker, 2000; Davenport and Klahr, 1998; Gibbert, Leibold, and Probst, 2002).

'CKM is the strategic process by which cutting-edge companies emancipate their customers from passive recipients of products and services, to empowerment as knowledge partners'; it is 'about gaining, sharing, and expanding the knowledge residing in customers, to both customer and corporate benefit' (Gibbert, Leibold, and Probst, 2002: 460). Gibbert and colleagues have identified the following five styles of CKM: prosumerism, mutual innovation, team-based co-learning, communities of practice, and joint intellectual property (IP) management. CKM constitutes a continuous strategic process by which companies

	KM	CRM	CKM
Knowledge sought in	Employee, team, company, network of companies.	Customer database.	Customer experience, creativity and (dis)satisfaction with products/services.
Axioms	'If only we knew what we know.'	'Retention is cheaper than acquisition.'	'If only we knew what our customers know.'
Rationale	Unlock and integrate employees' knowledge about customers, sales processes, and R & D.	Mining knowledge about the customer in company's databases.	Gaining knowledge directly from the customer, as well as sharing and expanding this knowledge.
Objectives	Efficiency gains, cost saving, and avoidance of re-inventing the wheel.	Customer base nurturing, maintaining company's customer base.	Collaboration with customers for joint value creation.
Metrics	Performance against budget.	Performance terms of customer satisfaction and loyalty.	Performance against competitors in innovation and growth, contribution to customer success.
Benefits	Customer satisfaction.	Customer retention.	Customer success, innovation, organizational learning.
Recipient of incentives	Employee.	Customer.	Customer.
Role of customer	Passive, recipient of product.	Captive, tied to product/ service by loyalty schemes.	Active partner in value-creation process.
Corporate role	Encourage employees to share their knowledge with their colleagues.	Build lasting relationships with customers	Emancipate customers from passive recipients of products to active co-creators of value.

Figure 4.2 CKM versus knowledge management and CRM (from Gibbert, Leibold, and Probst, 2002: 461)

enable their customers to move from passive information sources and recipients of products and services to empowered knowledge partners and empirical evidence points to CKM as a potentially powerful competitive tool, contributing to improved success of both companies and their customers (Gibbert, Leibold, and Probst, 2002: 467). Indeed, customer knowledge development inside the organization may affect in a positive way new product performance (Joshi and Sharma, 2004; cf. also Vicari and Cillo, 2006). To sum up, CKM incorporates principles of knowledge management and CRM, but 'moves decisively beyond both to a higher level of mutual value creation and performance' (Gibbert, Leibold, and Probst, 2002: 467).

But CRM still plays an important role as market-driven organizations 'develop intimate relationships with their customers, instead of seeing them as a means to a series of transactions' and these capabilities are 'built upon a shared knowledge base that is used to gather and disseminate knowledge about the market' (Day, 1999a: xi). Indeed, as buyer–seller relationships 'continue their transformation, a customer-linking capability – creating and managing close customer relationships – is becoming increasingly important' (Day, 1994a: 44). The customer-linking capability 'comprises the skills, abilities, and processes needed to achieve collaborative customer relationships so individual customer needs are quickly apparent to all functions and well-defined procedures are in place for responding to them' (Day, 1994a: 49). I therefore view CKM as one – from a knowledge-based perspective, essential – process within CRM. More will be said on relating in section 4.1.4.

Desouza and Awazu (2004, 2005a, 2005b) take a slightly different approach to CKM, as they propose three types of customer knowledge: (1) knowledge about the customer; (2) from the customer; and (3) to support the customer. Collectively, the three types of knowledge (about, to support, and from) make up the CKM construct (see Figure 4.3).

Knowledge about the customer is 'processed demographic, psychographic and behavioral information' (Desouza and Awazu, 2004: 12) and it is 'generated primarily through information processing activities' (Desouza and Awazu, 2005b: 119). Knowledge to support the customer 'is concerned with improving the user experience with products and services, which is critical for retaining customers' (Desouza and Awazu, 2004: 12; cf. also Davenport and Klahr, 1998). Managing knowledge that provides support for the customer requires an organization to leverage transaction data and information to personalize the pre-purchase, purchase, and post-purchase experiences, and ensuring a

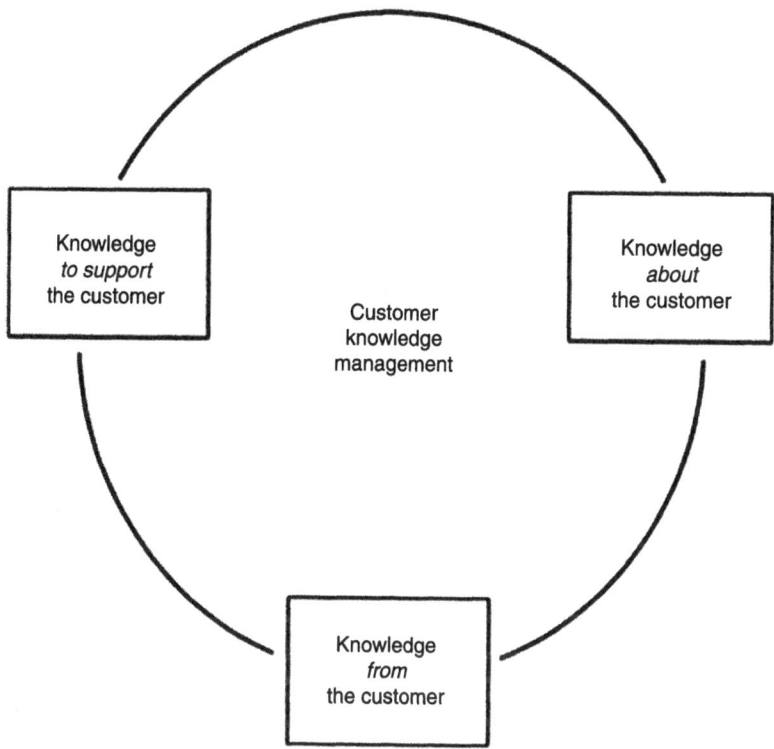

Figure 4.3 The CKM construct (from Desouza and Awazu, 2005b: 134)

pleasant user-experience is critical for retaining customers (Desouza and Awazu, 2005b: 124). Knowledge from the customers can be defined 'as the insights, ideas, thoughts, and information the organization receives from its customers' and these insights can be about current products and services, customer trends and future needs, and ideas for product innovations (Desouza and Awazu, 2005b: 130). Indeed, customers, in a sense, know products better than the organizations that produce them, which is why they represent a viable source of knowledge (Desouza and Awazu, 2004). This knowledge from the customer is concerned more with eliciting novel ideas and feedback (Desouza and Awazu, 2004, 2005b). Each of the dimensions of customer knowledge needs to be managed optimally and unless an organization can show competency in leveraging all three components, its CKM agenda will have an inherent weakness (Desouza and Awazu, 2005b: 134). It is important to note that the managing of knowledge from the customer has a greater human element than managing the other two types (knowledge about customers is almost completely leveraged via the use of technology, and knowledge to support customers has a balanced mix of technology and human components). Technology plays a support role in the management of knowledge from customers, while human interaction plays the primary role. An organization requires the human ability to comprehend the incoming knowledge about novel ideas and potential product innovation. In comparison with the other two types, knowledge from the customer is high in equivocality. Engaging such knowledge calls for a rich interaction between source and recipients. In pursuit of such valuable knowledge, most organizations try to promote rich human-to-human interactions (Desouza and Awazu, 2004, 2005b). Finally, one of the challenges for knowledge from customers is 'to integrate the various parts of the organization that need to use the knowledge', as the knowledge 'cannot simply stay with the marketing department' but 'must be shared with product engineering, research and development and the customer service department' (Desouza and Awazu, 2005a: 44). 'The trick is to capture knowledge gleaned from behavior or encounters unique to an individual customer and then reuse it by consolidating the findings and transferring the relevant pieces to other customers' (Vandermerwe, 2004: 34). The important issue of customer knowledge co-creation will be discussed in Chapter 4.2.3.

4.1.3.4   *Product development management (PDM)*

According to Natter, Mild, Feuerstein, Dorffner, and Taudes (2001: 1029), new product decisions 'have significant strategic implications that determine the future of a business and involve several functional areas within an organization', and Leonard (1998: 211) contends that '[o]ne of the most critical engines of renewal for companies is new-product development'. Indeed, the capability to bring products to market which comply with quality, cost, and development time goals is vital to the survival of firms in a competitive environment (Mild and Taudes, 2007). Some authors have shown that the resource-based view is well-suited to explain a firm's success in new product development (Dutta, Narasimhan, and Rajiv, 1999; Natter, Mild, Feuerstein, Dorffner, and Taudes 2001; Verona, 1999), and so is the knowledge-based view. New product development comprises knowledge creation and search and can be organized in different ways (Mild and Taudes, 2007) and according to Bell, Whitwell, and Lukas (2002: 82), product development is 'a particularly salient area for organizational learning inquiry for a number of reasons': it is often a team-based pursuit, it requires a high degree of interfunctional co-ordination, and it is frequently project-based. Indeed, there is a strong body of literature that deals with product development and product introduction from an organizational learning, knowledge management or market orientation perspective (cf. for example, Atuahene-Gima, 1996; Baker and Sinkula, 2005; Clark and Fujimoto, 1991; Dyck, Starke, Mischke, and Mauws, 2005; Hoegl and Schulze, 2005; Kusunoki, 2004; Kusunoki, Nonaka, and Nagata, 1998; Li and Calantone, 1998; Madhavan and Grover, 1998; Moorman, 1995; Schulze and Hoegl, 2006, to name but a few). Finally, product development is often difficult because 'the "need" information (what the customer wants) resides with the customer and the "solution" information (how to satisfy those needs) lies with the manufacturer' (Thomke, 2003: 244). Traditionally, 'the onus has been on manufacturers to collect the customer need information through various means, including market research and information gathered from the field', a process that 'can be costly and time-consuming because customer needs are often complex, subtle, and fast-changing' (ibid.). But here the ultimate aim is to progress from idea generation to launch only those products for which success is guaranteed, and as the firm moves through the process, at each stage knowledge is acquired and evaluated about whether the product under development should be progressed or terminated (Chaston, 2004: 160).

Although the classic view of entry point for the process model is idea generation (Chaston, 2004), Li and Calantone (1998) posit that certain market and internal competencies are key antecedent determinants of success at the idea generation stage. The greater the firm's knowledge of customer needs, the more probable it is that new ideas will be generated that offer the greatest potential for market success (Chaston, 2004: 160).

Nonaka and associates (for example, Imai, Nonaka, and Takeuchi, 1985; Takeuchi and Nonaka, 1986) have already discussed the issues of creating and transferring knowledge in product development projects more than twenty years ago and the theory of organizational knowledge creation is thoroughly grounded in and backed up by empirical research on such projects (for example, Dyck, Starke, Mischke, and Mauws, 2005; Hoegl and Schulze, 2005; Nonaka and Takeuchi, 1995; Nonaka, Byosiere, Borucki, and Konno, 1994; Schulze and Hoegl, 2006). In fact, even though many vital processes of innovation, change, and renewal in organizations can be analysed through the lens of knowledge conversion (Nonaka, von Krogh, and Voelpel, 2006), knowledge creation and transfer in product development projects seem to be particularly important, as the research focus by both Western (cf., for example, Leonard, 1998; Leonard and Sensiper, 1998; Leonard-Barton, 1992; von Hippel, 1994) and Japanese scholars (cf., for example, Aoshima, 2002; Cusumano and Nobeoka, 1998; Kusunoki, Nonaka, and Nagata, 1998; Nobeoka, 1995; Nobeoka and Cusumano, 1997) has also shown. Indeed, 'the ability to import knowledge from the market' is a principal component of the product development process (Leonard, 1998: 179). Baba and Nobeoka (1998) in their study on the introduction of 3-D CAD systems, even speak of 'knowledge-based product development'. Moreover, Nonaka, Kohlbacher, and Holden (2006) suggest that members of a product development project share ideas and viewpoints on their product design in a *ba* that allows common interpretation of technical data, evolving rules of thumb, an emerging sense of product quality, effective communication of hunches or concerns, and so on.

As we will also see in the Mazda case study in Chapter 5.6, developing, disseminating, and implementing a unique concept is an essential step in product development. Natter, Mild, Taudes, and Geberth (2004: 472) define a product concept as 'a description of a product in accordance with attributes perceived by the target customers'. However, these concepts usually tend to be highly tacit and as such difficult to transfer to others. Indeed, if conceptual marketing is used for concept development, 'then a number of product ideas will be evaluated on the basis of tacit knowledge gained only through market involvement' (Natter, Mild, Taudes, and Geberth, 2004: 472). Therefore, especially in the concept development stage, it is critical to articulate images rooted in tacit knowledge and meaningful information arises as a result of the conversion of tacit knowledge into articulable knowledge (Nonaka, 1990a). Explaining the process of externalization of tacit knowledge into explicit knowledge, Nonaka and Takeuchi (1995: 64–5) maintain that a 'frequently used method to create a concept is to combine deduction and induction' and highlight the example of Mazda, which combined these two reasoning methods when it developed the new RX-7 concept. In fact, as Nonaka and Toyama (2003) argue, abduction or retroduction might be even more effective than induction or deduction to make a hidden concept or mechanism explicit out of accumulated tacit knowledge.

As mentioned in the discussion of market research (4.1.3.2), the way of capturing customer needs and translating them into a product concept has been termed 'empathic design' (Leonard-Barton, 1991;

Leonard, 1998; Leonard and Rayport, 1997; Leonard and Swap, 2005b). Leonard (1998: 194, emphasis removed), defines empathic design as 'the creation of product or service concepts based on a deep (empathetic) understanding of unarticulated user needs'. It is 'a set of techniques, a process of developing deep empathy for another's point of view and using that perspective to stimulate novel design concepts' (Leonard and Swap, 2005b: 82). Empathic design differs from contextual inquiry precisely because it does not rely on inquiry; in the situations in which empathic design is most useful, inquiry is useless or ineffective (Leonard, 1998: 288n). The more deeply a researcher can get into the mindset, the perspective, of a prospective or actual user, the more valuable is the knowledge thus generated (Leonard, 2006).

Obviously, the knowledge gained and generated through market research, empathic design, product development, and so on, should not vanish after the project finishes. Indeed, it is essential to retain vital knowledge and share and transfer across functions, between projects, as well as generations of projects and products (for example, Cusumano and Nobeoka, 1998; Nobeoka, 1995; Nobeoka and Cusumano, 1997). As Cusumano and Nobeoka (1998: 175) put it: 'In addition to overlapping projects and using cross-functional teams, companies have various organizational and technological mechanisms to help them capture knowledge about designs or manufacturing processes and then transfer this knowledge across different projects or different generations of products.' In fact, successful new product development at least partially depends on the ability to understand technical and market knowledge embodied in existing products, and the adaptation of this knowledge to support new product development (Aoshima, 2002; Iansiti, 1997; Iansiti and Clark, 1994).

Aoshima (1996, 2002) studied the transfer of knowledge across different projects in the automotive industry and examined two types of knowledge. One related to the development of specific components and the other related to integration of different components. Aoshima (1996; 2002) called the former 'local' knowledge and the latter 'system' or 'integrative' knowledge. For component knowledge, he found that archival-based mechanisms, such as documents, reports, written engineering standards, and computerized tools, were more effective in promoting knowledge retention than individual-based mechanisms such as transfer of people or direct communication between members of different projects. This seems to be because component-level knowledge is rather specialized and can be written down. For system or integrative knowledge, however, he found that companies did better if they relied more on individual-based mechanisms, primarily face-to-face communication and transfer of people from one project to another. This is probably because this kind of knowledge is difficult to communicate and write down (Aoshima, 1996, 2002; cf. also Cusumano and Nobeoka, 1998). These findings also seem to be consistent with von Hippel (1994).

As discussed in Chapter 3.4, Nonaka's publications (for example, Nonaka, 1994; Nonaka and Takeuchi, 1995) have drawn attention to Japanese firms as knowledge-creating companies, and the difference, it was argued, between Japanese and Western firms lies in the focus on tacit knowledge of the former and explicit knowledge of the latter (Hedlund and Nonaka, 1993; Takeuchi and Nonaka, 2000). Additionally, the practices of the Japanese 'knowledge-creating company' are also interesting from a marketing perspective, 'because they demonstrate how companies mobilize all employees to learn more about markets and how to captivate customers' (Johansson and Nonaka, 1996: 164). As a matter of interest, since its beginning, the theory of corporate knowledge creation has been closely related to the field of marketing due to its focus on new product development projects (Nonaka, 1991; Takeuchi and Nonaka, 1986). The same is also true for Leonard's (1998) work on knowledge assets. Marketing's first detailed glimpse of Japanese firms' knowledge-creating capabilities came with the publication in 1995 of Nonaka and Takeuchi's book *The Knowledge-Creating Company*. Indeed, the fact that creating, sharing, and managing (marketing) knowledge are particularly crucial in new product development projects has frequently been recognized and discussed (cf., for example, Bell, Whitwell, and Lukas, 2002; Hoegl and Schulze, 2005; Madhavan and Grover, 1998; Moenaert and Souder, 1990; Schulze and Hoegl, 2006). Moreover, as marketing organizations serve as corporate links between customers and their organization's manufacturing and R&D operations (Riesenberger, 1998), the integration of and knowledge exchange between R&D and marketing have also been treated as important issues (for example, Griffin and Hauser, 1996; Song and Parry, 1993). When organizations remove the functional barriers that impede the flow of information from development to manufacturing to sales and marketing, they improve the organization's ability to make rapid decisions and execute them effectively (Slater and Narver, 1995: 65; cf. also Clark and Fujimoto, 1991; Nonaka, 1990a). Indeed, '[n]ew product introduction in subsidiaries around the world can benefit from knowledge management systems that address the needs of best marketing practices and stimulate cross-subsidiary learning through access to information and knowledge exchange among employees' (Riesenberger, 1998: 101).

4.1.4 The long road to marketing knowledge and knowledge-based marketing

The literature reviews on market orientation and organizational learning (4.1.1), marketing knowledge (4.1.2), and areas of marketing knowledge creation and application (4.1.3) have shown that there are many research streams contributing to the field and that there are many different angles, processes, and functions from which one can take a knowledge-based view of marketing. Despite different approaches to and conceptualizations of market orientation (cf., for example, Figure 4.1), the different approaches have much in common and build upon and draw from each other. As a result, the whole research stream of market orientation has a comparatively high degree of consistency and homogeneity. In contrast to that, the treatment of market and specifically marketing knowledge (including customer knowledge) does not seem to have such a common ground unifying scholars. Other areas such as SCM, market research, CRM (including CKM, DCM, and the lead user concept), and PDM have contributed to the field, or at least have the potential to do so. But most of the time, they have done so independently of each other and there is no comprehensive framework integrating the different approaches and areas of research.

Figure 4.4 gives an overview of the related research streams, concepts, and antecedents of a knowledge-based view of marketing which have been discussed in Chapters 3 and 4.1, and illustrates their theoretical/conceptual affiliations and interrelations. It emphasizes once again how scattered and heterogeneous the state-of-the-field of knowledge-based marketing is and how many different streams of research have to be considered. The dashed line above knowledge-based marketing is meant to emphasize that although all of the above research streams and areas contribute, there is no theory that links and integrates the different approaches to and into knowledge-based marketing. Differences between organizational learning and knowledge management and the resource-based and the knowledge-based view of the firm have been touched upon in Chapter 3 and will be highlighted again below.

In addition to the semantic uncertainties associated with the notion of marketing knowledge (cf. above, 4.1.2), the literature review reveals scholars' assumptions or suppositions about the role of knowledge in marketing as well as organizational learning and knowledge management in marketing. These can be summarized as follows:

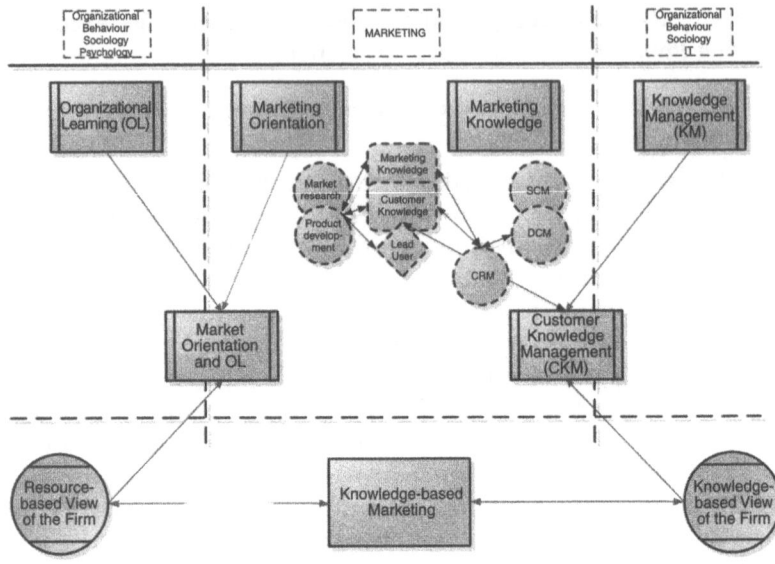

Figure 4.4 Related concepts and antecedents of knowledge-based marketing (author's own illustration)

(1) There does not seem to exist a knowledge-based view of marketing, let alone a knowledge-based marketing theory. As shown in Chapter 4.1.1 on market orientation, many authors refer to, or draw from, the resource-based view of the firm, but none builds on the knowledge-based view of the firm.

(2) There does not seem to exist a knowledge management approach to marketing, that is, an application of knowledge management concepts to marketing. As shown in Chapter 4.1.1 on market orientation, many authors refer to, or draw from, the organizational learning literatures, but none builds on knowledge management research.

(3) The terms 'information' and 'knowledge' seem to be used interchangeably and indiscriminately. As shown in Chapter 3 – specifically sections 3.1 and 3.4.1 – knowledge goes far beyond the concept of information and a distinction between tacit and explicit knowledge is crucial for the analysis und understanding of the organizational knowledge-creation process and innovation.

(4) The notion of marketing knowledge is closely bound up with the inputs derived from various sources: academic marketing knowledge and market research, and applied to marketing research with special reference to customers and new product development.

Schlegelmilch and Penz (2002) and Achrol and Kotler (1999) are two of the few exceptions to grasp the importance of marketing knowledge creation.

(5) There is a good deal of hypothesizing about marketing as a knowledge-based activity, but a dearth of studies about how firms consciously develop and apply a knowledge-based approach.

(6) The literature is correspondingly short of case material, the only exception being Chaston (2004).

(7) The literature to date is completely silent about the application of a knowledge-based approach to marketing as in explicitly international pace cross-cultural contexts.

(8) With the exception of Chaston and fellow researchers (Chaston, 2004; Chaston, Badger, Mangles, and Sadler-Smith, 2001, 2003; Chaston, Badger, and Sadler-Smith, 2000, 2001), the discussion focuses on large organizations, implying that a knowledge-based approach to marketing is less relevant to smaller ones.

How the knowledge-based view of the firm overcomes the main shortcomings and limitations has been highlighted in Chapter 3.3: the resource-based view treats knowledge as one resource, but empirical and theoretical research on the resource-based view of the firm so far has been mainly focused on how firms keep their unique resources and resulting competitive advantages through such conditions as imperfect substitutability and limited mobility of resources (Nonaka, Toyama, and Nagata, 2000: 7–8; Nonaka and Toyama, 2003: 4; cf. also Amit and Schoemaker, 1993; Barney, 1991; Nonaka, von Krogh, and Voelpel, 2006; Peteraf, 1993; Wernerfelt, 1984). Therefore, Nonaka and Toyama (2003: 4) conclude that – although it deals with the dynamic capability of the firm – 'the resource-based view of the firm fails to explain the dynamism in which the firm continuously builds such resources through interactions with the environment' (cf. also Nonaka, Toyama, and Nagata, 2000: 7). 'What is missing in the resource-based approach is a comprehensive framework that shows how various parts within the organization interact with each other over time to create something new and unique' (Nonaka and Takeuchi, 1995: 49). I propose that this same limitation applies to market orientation and organizational learning as discussed in Chapter 4.1.1.2. This research stream obviously deals with dynamic capabilities of firms – marketing capability, marketing information processing capabilities, and so on – but it fails to show how various parts within – and actually also outside – the organization interact to create new knowledge and leverage innovation.

Kyriakopoulos and Moorman (2004: 224) contend that although market orientation seems to be closely related to exploitation and exploration (to be discussed in greater detail below, 4.2.2) they are distinct concepts for several reasons. First, market orientation is a firm-level trait whereas exploitation and exploration are project-level strategies, which is why they argue that a firm's market orientation creates the context within which project-level marketing strategies can cross-pollinate (Kyriakopoulos and Moorman, 2004: 224). Consistent with this, Slater and Narver (1995), as will be recalled, argue that a market orientation provides norms for learning from customers and competitors but that it is distinct from adaptive or generative learning strategies. Second, none of the current views of market or customer orientation is implicitly exploitation- or exploration-focused. Therefore, this element must be accounted for by additional strategic factors in the firm (Kyriakopoulos and Moorman, 2004; see also Day, 1999a; Slater and Narver, 1999).

Nonaka and Takeuchi (1995: 45) – while acknowledging 'the affinity with [their] own thinking' – identify some critical limitations often found in the literature on organizational learning. First, as seen in Senge (1990), organizational learning theories basically lack 'the view that knowledge development constitutes learning' (Weick, 1991: 122). Most of them are trapped in a behavioural concept of 'stimulus-response'. Second, most of them still use the metaphor of individual learning (Dodgson, 1993; Weick, 1991). In the accumulation of over twenty years of studies, they have not developed a comprehensive view on what constitutes 'organizational' learning. Third, there is widespread agreement that organizational learning is an adaptive change process that is influenced by past experience, focuses on developing or modifying routines, and is supported by organizational memory. As a result, the theories fail to conceive an idea of knowledge creation (even though there are exceptions). The fourth limitation is related to the concept of 'double-loop learning' (Argyris, 1977) or 'unlearning' (Hedberg, 1981) as well as to a strong orientation towards organizational development. Following the development of Argyris and Schön's (1978) theory of organizational learning, it has been widely assumed implicitly or explicitly that double-loop learning – the questioning and rebuilding of existing perspectives, interpretation frameworks, or decision premises – can be very difficult for organizations to implement by themselves. In order to overcome this difficulty, the learning theorists argue that some kind of artificial intervention, such as the use of an organizational development programme, is required.

The limitation of this argument is that it assumes that someone inside or outside an organization 'objectively' knows the right time and method for putting double-loop learning into practice (Nonaka and Takeuchi, 1995: 46). 'Seen from the vantage point of organizational knowledge creation, double-loop learning is not a special, difficult task but a daily activity for the organization', as organizations 'continuously create new knowledge by reconstructing existing perspectives, frameworks, or premises on a daily basis' (Nonaka and Takeuchi, 1995: 46). 'In other words, the capacity for double-loop learning is built into the knowledge-creating organization without the unrealistic assumption of the existence of a "right" answer' (ibid.).

All of these limitations identified by Nonaka and Takeuchi (1995) basically also apply to the literatures in the field of market orientation, as they build upon the theories of organizational learning. In fact, a knowledge-based theory of marketing cannot merely deal with passive learning processes, but instead needs to embrace active knowledge creation and management. Nonaka and Takeuchi (1995: 49, original emphasis) have also pointed out that 'there are very few studies on how knowledge is created within and between business organizations' because '[a]t the core of concern of these theories is the acquisition, accumulation, and utilization of *existing* knowledge', while they lack the perspective of 'creating new knowledge'. A quick look again at Chapter 4.1.1.2 immediately confirms this problem also for the field of marketing. Indeed, as we have seen above, Sinkula (1994) – based on Huber's (1991: 90) four organizational learning-related constructs: knowledge acquisition, information distribution, information interpretation, and organizational memory – depicts market information processing (that is, organizational learning in marketing) as a four-stage process that includes information acquisition, information dissemination, shared interpretations, and storage (organizational memory) (cf. also Slater and Narver, 1995). He further proffers that market information processing 'is a function of what the organization has learned in terms of both facts about its relevant markets and its particular way of acquiring, distributing, interpreting, and storing information' (Sinkula, 1994: 37). The need for a knowledge-based marketing theory is obvious.

Both the shortage of research on organizational creation of marketing knowledge and the dearth of studies about how firms consciously develop and apply a knowledge-based approach as well as the corresponding shortage of case material (Kohlbacher, Holden, Glisby, and Numic, 2007) – Chaston (2004) has been a long-awaited exception

– have prompted the empirical research project described in Chapters 2 and the Appendix, and this book aims to contribute to closing this disconcerting gap in the marketing and knowledge management literature. According to Chaston (2004: 155), '[i]n a world where other firms are seeking to expand their market share, successful firms often can only stay ahead of the competition by exploiting new knowledge to offer improved products or processes that deliver new forms of added value to their customers'. Nevertheless, the literature to date is completely silent about the application of a knowledge-based approach to marketing as in explicitly international pace cross-cultural contexts. In fact, there is a general shortage of research and academic writing on marketing knowledge and marketing from a knowledge-based view. The market orientation literature (4.1.1), specifically the works of Jaworksi and Kohli (Jaworski and Kohli, 1993; Jaworski, Kohli, and Sahay, 2000; Kohli and Jaworski, 1990; Kohli, Jaworski, and Kumar, 1993), Narver and Slater (Narver and Slater, 1990; Narver, Slater, and Maclachlan, 2004; Slater and Narver, 1995; Slater and Narver, 1994, 1998, 1999), and Day (Day, 1990, 1994a, 1994b, 1998, 1999a, 1999b; Day and Montgomery, 1999), can certainly be seen as attempts to deal with marketing knowledge processes in firms. But the limitations discussed above apply. Apart from this literature in the English language, there appears to be no article in the German-language and Japanese-language management literature which discusses knowledge-based approaches to marketing. In German, the only contribution approaching marketing and knowledge management was about market research as a knowledge management function, stating that marketing research 'has a key position within the market related knowledge management' (Grundei, 2000: 342). As for Japanese, there only seem to be the works by Furukawa (most prominently Furukawa, 1999a, 1999b; Katahira, Furukawa, and Abe, 2003) and Ogawa (2000). The former use Nonaka and Takeuchi's (1995) SECI model and the concept of *ba* (Nonaka and Konno, 1998) to analyse and explain how knowledge about products is shared both between consumers – and thus leads to the diffusion and adoption of innovations – and consumers and companies – which contributes to the literature on customer knowledge and market research.

I strongly believe that in an increasingly global business environment, the creation and transfer of marketing knowledge and intra-firm collaboration through knowledge-based approaches to marketing will become more and more crucial as a determinant for corporate competitive advantage and survival of firms. Indeed, as marketing affairs are one of the most knowledge-intensive aspects of a company, applying knowledge management concepts and practices to the field of marketing and to marketing functions will prove especially efficient and effective. But as the above review and discussion have shown, marketing lacks a knowledge-based framework in order to analyse and explain marketing-related knowledge processes in firms. Basically, only the research by Schlegelmilch and fellow researchers (Schlegelmilch, Ambos, and Chini, 2003; Schlegelmilch and Chini, 2003; Schlegelmilch and Penz, 2002) can be seen as an exception to this shortage. This book aims to contribute to closing these disconcerting gaps and to overcome misconceptions by presenting findings from a recent empirical study and by analysing various case studies, revealing these firms' strength and ability for creating and leveraging (marketing) knowledge both locally and globally (see Chapters 5 and 6.1). Finally, as Gummesson (2001: 29, added emphasis) notes, '[s]ervices and B-to-B (business-to-business) marketing, relationships, networks, quality, *knowledge management*, brand equity, green marketing, information technology and other developments have had some impact but have not made marketing theorists bake a cake according to a new recipe, just to add decorations on the glazing of the old cake'. The next section (4.2) is therefore an attempt to engage in the first steps of building a knowledge-based framework for marketing and to build theory. I have been obliged to use – or rather create – a new recipe but the challenge will be to successfully bake the new cake.

4.2 Knowledge-based approaches to marketing

> The bottom line is that markets are changing faster than our marketing. The classic marketing model needs to be future-fitted. Marketing must be deconstructed, redefined, and stretched. (Kotler, Jain, and Maesincee, 2002: x)

'At an organisational level in a modern economy knowledge is the most important resource within the company' (Chaston, 2004: 2). Both empirical as well as literature research have shown a tendency for the creation and transfer of (marketing) knowledge and intra-firm collaboration through knowledge-based marketing to become more and more crucial as a determinant for corporate competitive advantage and survival of firms in an increasingly global business environment. As stated above, marketing affairs are highly knowledge-intensive, applying knowledge management concepts and practices to the field of marketing and to marketing functions might prove especially efficient and effective. As large parts of marketing knowledge are tacit and hard to codify, face-to-face communication and the integration of local staff into marketing processes and decision-making will be a critical factor for global marketing knowledge sharing that leads to successful marketing and sales achievements.

However, as has been mentioned above, despite the growing recognition of the need for knowledge-based approaches to marketing, there seem to be only a few pioneer firms that are already taking or trying to take such an approach. The case studies to be presented in Chapter 5 will show how these firms face the challenge of an increasingly global business environment with fierce competition and take up and master the challenge with the help of knowledge-based marketing. As Hansen and Nohria (2004: 22) correctly note, the ways for MNCs to compete successfully by exploiting scale and scope economies or by taking advantage of imperfections in the world's goods, labour and capital markets are no longer as profitable as they once were, and as a result, 'the new economies of scope are based on the ability of business units, subsidiaries and functional departments within the company to collaborate successfully by sharing knowledge and jointly developing new products and services'. In fact, this statement strongly supports the need for knowledge-based (approaches to) marketing.

Before moving on to the discussion of marketing knowledge and knowledge-based marketing, it is necessary to clarify what is meant by marketing, that is, what is the definition of marketing underlying this book. In 2004, the American Marketing Association (AMA) announced a new definition of marketing:

> Marketing is an organizational function and a set of processes for creating, communicating, and delivering value to customers and for managing customer relationships in ways that benefit the organization and its stakeholders. (American Marketing Association, 2004)

Note how this definition focuses on delivering value to customers and on the management of relationships, and includes all stakeholders of the firm. In a similar vein, Gummesson (2003b: 168) defines marketing as 'interaction in networks of commercial relationships'. When exploring the role of relationships, networks, and the business ecosystem in Chapter 4.2.2.2 and the notion of marketing knowledge co-creation in Chapter 4.2.3, the significance of these definitions will become even more obvious.

4.2.1 Marketing knowledge

As shown in Chapter 4.1, it may not be outlandish to suggest that the marketing discipline is tying itself up in semantic knots while it struggles to create consensus on an agreed definition of the term 'marketing knowledge', its practical scope, and supporting constructs. In fact, the current marketing literature does not offer one satisfactory definition of marketing knowledge which is also amenable to the investigation of international marketing interactions (Kohlbacher, Holden, Glisby, and Numic, 2007). However, providing a clear definition of the term 'marketing knowledge' is absolutely essential for the development of a knowledge-based theory of marketing and for any discussion of knowledge-based marketing.

Wierenga (2002: 355) claims that 'restricting marketing knowledge to academic knowledge is unnecessary and not productive' and that '[m]arketing decision-makers in practice have a much richer treasure of marketing knowledge at their disposal than the "codified body of knowledge" that has emerged from systematic academic research'. Based on the insight that marketing knowledge can be deep knowledge or surface knowledge, explicit or tacit knowledge, and objective or subjective knowledge (Wierenga and van Bruggen, 2000), Wierenga (2002: 356, removed emphasis) defines marketing knowledge as '[a]ll the insights and convictions about marketing phenomena that marketing managers use or can use for making marketing decisions'. While academic marketing knowledge 'is characterized by terms such as marketing laws, marketing principles, empirical generalizations, and marketing science', marketing practitioners 'use much more knowledge than only the products of marketing science', as they 'usually have extensive *experience*, which produces a significant amount of *expertise*' (Wierenga, 2002: 356, 357, original emphasis). Grønhaug (2002: 370–1) deals with academic marketing knowledge as produced, taught, and disseminated by marketing academicians in greater detail and contends that 'marketing knowledge should be helpful to businesses in understanding their customers and business environments, allowing business firms to make wise decisions, take successful actions and thus keep their competitive edge'. He further reminds us that if 'marketing knowledge is to yield competitive advantage, it must also be superior, probably developed to the degree of expert knowledge' (Grønhaug, 2002: 371).

I strongly support the call for including the marketing knowledge that managers use for decision-making in the concept of marketing knowledge (see also below, holistic marketing knowledge). However, as 'it seems a safe statement that academic marketing knowledge has a modest share in the total quantity of marketing knowledge that a marketing manager uses' (Wierenga, 2002: 359), I do not consider this dichotomy very helpful. Rather I propose different types of marketing knowledge according to the different entities that carry the knowledge or whom the knowledge is about. In fact, customer knowledge as a critically important resource has already been highlighted in Chapter 4.1.3.3. Besides, rather than there being objective, generalizable marketing knowledge, it is probably firm-specific and therefore accumulated at the level of the individual firm (cf., for example, Kohli and Jaworski, 1990; McIntyre and Sutherland, 2002; Menon and Varadarajan, 1992). Indeed, because firms are 'embedded in hostile, ever-changing environments, knowledge about the actual context will (in most cases) also be needed in addition to the general marketing knowledge' (Grønhaug, 2002: 371).

In addition to being divided into functional areas, organizations all possess domains of knowledge. Domains of knowledge are 'areas of distinct knowledge about certain things that have a common theme' (Grieves, 2006: 57). The most common domains of knowledge in an organization are knowledge about products, knowledge about customers, knowledge about employees, and knowledge about suppliers. Knowledge about products deals with all the information an organization has about a product: how it needs to be designed, how it needs to be manufactured, the functionality it needs to have, and so on. Knowledge about customers deals with customer-specific knowledge: their requirements, their procedures for doing business, their ways of making decisions. Knowledge about employees deals with their knowledge of areas of expertise: the processes they perform, their expertise in certain areas. Finally, knowledge about suppliers deals with the expertise of suppliers: the products that suppliers have to offer, their manner of doing business, their quality of work, their reliability, and so on (Grieves, 2006: 57).

Trusting that no marketing scholar or practitioner is likely to disagree on marketing affairs being one of the most knowledge-intensive parts of a company, I propose the following definition of marketing knowledge:

Marketing knowledge is all knowledge, both declarative as well as procedural, concerning marketing thinking and behaviour in a corporation.

Obviously, this leads to a very broad concept of marketing knowledge, but given the early stage of research on knowledge-based approaches to marketing, this definition proves to be a helpful guidance for the exploratory empirical study (cf. Chapters 5, 6 and Appendix). Indeed, Bjerre and Sharma's (2003: 140) pronouncement to the effect that the 'important thing is not one specific piece of knowledge, but an entire package that includes knowledge about clients, competitors, local institutions, suppliers etc.' underscores the importance of such a comprehensive concept of marketing knowledge. As Pollard (2006: 21) reminds us, 'marketing knowledge of a company develops both in-house and through external contact', another feature of holistic marketing knowledge that comprises knowledge of and about other entities and stakeholders in the market place and the ecological system of a firm. In fact, successful companies 'create collaborative networks to gain and disseminate knowledge' (Kotler, Jain, and Maesincee, 2002: 113). Therefore, the above definition includes both tacit as well as explicit knowledge about products, markets, customers, competitors, partners, marketing processes, and marketing strategy. Finally, it includes also experiences of past marketing efforts such as new product introductions, as well as future expectations. Note that a finer and narrower definition of marketing knowledge leads to the definition of one of these subunits of marketing knowledge, such as customer knowledge, competitor knowledge, and so on. Marketing knowledge itself is a holistic concept and has deliberately been defined in a broad way.

Declarative and procedural knowledge have been defined in Chapter 3.1. Besides, as will be recalled (cf. 4.1.2.4), for Rossiter (2001: 10), 'marketing knowledge is declarative ("know what") and ... exists independently of, and should be distinguished from, marketing skills or procedural knowledge ("know-how")'. In my holistic definition of marketing knowledge, the term includes both aspects: declarative marketing knowledge as knowledge about facts, stakeholders, the environment , and so on, relevant for marketing affairs of a firm, and procedural marketing knowledge as the knowledge of marketing processes and the know-how to process this knowledge as well as knowledge from stakeholders, for example. Indeed, very importantly, competitor, customer, partner, and supplier knowledge not only means knowledge *about* competitors, customers, partners, and suppliers, but also knowledge *from* competitors, customers, partners and suppliers. Gibbert, Leibold, and Probst (2002) and Desouza and Awazu (2004, 2005b) have stressed this in the case of customer knowledge already (cf. 4.1.3.3) and the importance of knowledge from alliance partners (competitors, suppliers, partners) and suppliers was discussed in Chapters 3.7 and 4.1.3.1. Finally, the explicit mentioning of both declarative and procedural knowledge is also meant to take account of the fact that marketing is defined as both an organizational function and as a set of processes (American Marketing Association, 2004; see also above).

Furthermore, the importance of tacit knowledge and its relevance should have become clear in Chapter 3 and Chapter 4.1.2.4, and will be further illustrated in the stories about the informant companies and my analysis of their experiences in Chapters 5 and 6.1. Indeed, '[m]ost of the knowledge that makes an organization competitive is its tacit, not its explicit, knowledge' (Dixon, 2000: 96). Given the significance of tacit knowledge, it is important to note that the expression 'all knowledge' in the above definition of marketing knowledge actually refers to a concept I would like to term 'holistic knowledge'. This 'holistic marketing knowledge' is a combination and synthesis of both tacit and explicit marketing knowledge (cf. Figure 4.5). The term 'holistic knowledge' will be dealt with in greater detail in section 4.2.3. In a sense, it is also similar to the concept of 'common knowledge' proposed by Dixon (2000: 11) who defines it as 'the knowledge that employees learn from doing the organization's tasks' (cf. Chapter 3.1). Note that common knowledge 'is always linked to action', as it is 'derived from action and it carries the potential for others to use it to take action' (Dixon, 2000: 13). Figure 4.5 also shows that (holistic) marketing knowledge is inherently tacit at its core and that the explicit knowledge around it might be but the tip of the iceberg of the real value of the knowledge in question. Besides, all types of marketing knowledge – for example, customer knowledge, product knowledge, and so on – have both tacit and explicit components, which need to be considered when taking a holistic perspective of marketing knowledge. According to Kotler, Jain, and Maesincee (2002: 113), knowledge is 'information that has been edited, put into context, and analysed in a way that makes it meaningful'. Obviously, Nonaka's (1994) theory of organizational knowledge creation and his SECI model play an important role here (cf. 3.4 and 4.2.3).

Last, but not least, marketing knowledge is an organizational resource that reduces the effects of ambiguity and complexity in cross-border interactions (Simonin, 1999a), as well as a 'primary source of co-ordinating power' in business networks (Achrol and Kotler, 1999: 157). In fact, according to Hanvanich, Dröge, and Calantone (2003: 126), 'marketing knowledge should enable firms to identify competent business partners so as to build capabilities'. But '[to] understand the complex nature of the marketing knowledge more frequent and prolonged direct, face-to-face contact between firms is needed' (Bjerre and Sharma, 2003: 140). This is confirmed by Madhavan and Grover

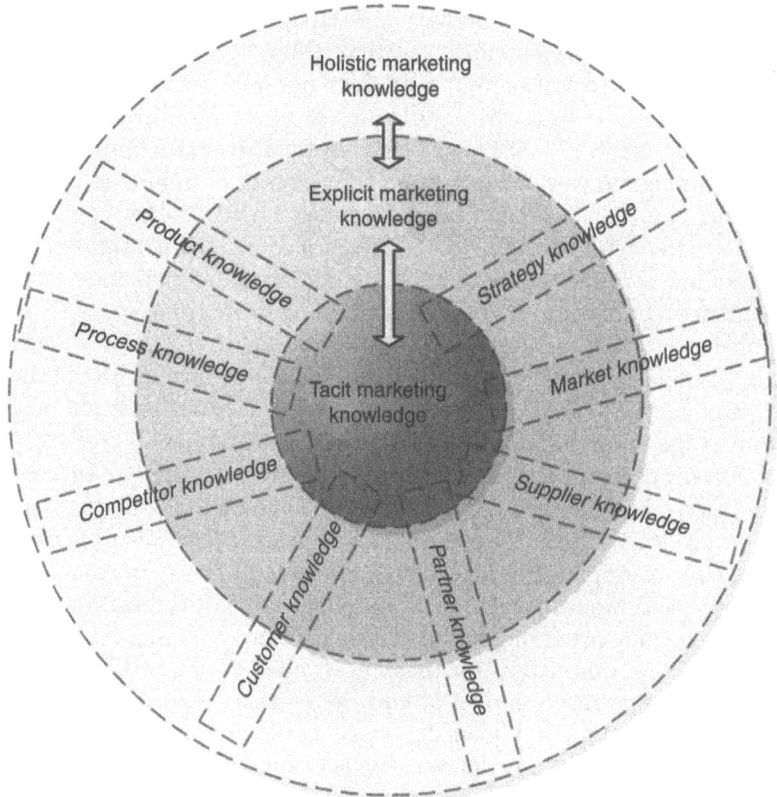

Figure 4.5 A holistic notion of marketing knowledge (author's own illustration)

(1998: 6) who found that rich personal interaction directly affects the efficiency and effectiveness with which embedded knowledge is converted to embodied knowledge.

Having clarified what marketing knowledge is – an issue that had been pending for a long time – it is now time to turn to the concept of 'knowledge-based marketing', which builds on marketing knowledge as its key resource.

4.2.2 Knowledge-based marketing

Knowledge is no good if you don't apply it. (Goethe)

As mentioned in Chapter 3, '[v]irtually all business conducted today is global business' (Thomas, 2002: 3); national economies have become increasingly deregulated and have opened up opportunities for international trade and competition so that it has 'become the norm for organizations to compete for market share not only with their national competitors but also with international ones' (Trompenaars and Woolliams, 2004: 27). Besides, in such 'an era of ever faster innovation cycles combined with an increasing convergence of industries ... and intense and global competition, advantages tend to erode quickly' (Ambos and Schlegelmilch, 2005: 23). As a result, the majority of all marketing activities will have to be international and marketing managers need to develop a deep understanding of the idiosyncrasies of global marketing (Schlegelmilch and Sinkovics, 1998). At the same time, scholars and practitioners around the globe have identified the capability of MNCs to create and efficiently transfer and combine knowledge from different locations worldwide as an increasingly important determinant of competitive advantage, corporate success, and survival (cf., for example, Asakawa and Lehrer, 2003; Bartlett and Ghoshal, 2002; Chini, 2004; Desouza and Awazu, 2005b; Doz, Santos, and Williamson, 2001; Gupta and Govindarajan, 2000a; Macharzina, Oesterle, and Brodel, 2001; Schulz and Jobe, 2001). It is therefore high time to include marketing in the knowledge management agenda.

This section summarizes the theoretical and empirical insights into a comprehensive macro-model of knowledge-based marketing. First, a holistic definition of knowledge-based marketing is provided and explained. Second, the key players and actors in knowledge-based marketing are discussed and the main influencing factors of knowledge-based marketing are presented.

4.2.2.1 *Definition*

A key issue in the literature on organizational learning and knowledge management is how successfully firms learn when they are exploiting current knowledge and skills versus exploring new knowledge and skills, and a long tradition of research suggests that these are competing strategies (Kyriakopoulos and Moorman, 2004; March, 1991; Miller, Zhao, and Calantone, 2006). But this view has also been challenged, arguing that firms must engage in both strategies (for example, He and Wong, 2004; Jansen, Van den Bosch, and Volberda, 2005; Kyriakopoulos and Moorman, 2004; Levinthal and March, 1993; Lewin and Volberda, 1999). Levinthal and March (1993: 105) put it like this:

An organization that engages exclusively in exploration will ordinarily suffer from the fact that it never gains the returns of its knowledge. An organization that engages exclusively in exploitation will ordinarily suffer from obsolescence. The basic problem confronting an organization is to engage in sufficient exploitation to ensure its current viability and, at the same time, to devote enough energy to exploration to ensure its future viability. Survival requires a balance, and the precise mix of exploitation and exploration that is optimal is hard to specify.

Kyriakopoulos and Moorman (2004) identified research in various fields that has recently shifted focus from whether to how firms can achieve a complementarity of the exploitation and exploration strategies: Brown and Eisenhardt (1997), for example, introduce semi-structured and time-paced strategies as managerial tools to achieve this dynamic balance in product innovation. Likewise, the integration of exploration and exploitation is central to work examining dynamic or combinative capabilities (Grant, 1996a; Kogut and Zander, 1992; Teece, Pisano, and Shuen, 1997). In the product development literature, scholars often study the degree of fit between a new product and prior activities (for example, marketing and technological synergy: Henard and Szymanski, 2001; Montoya-Weiss and Calantone, 1994; Moorman and Miner, 1997; Song and Parry, 1997). Kyriakopoulos and Moorman (2004: 220) contribute to this literature by suggesting that a firm's market orientation can systematically promote synergies between exploratory and exploitative marketing strategy activities because 'a firm's market orientation reduces the tensions between exploration and exploitation strategies and creates the opportunity for cross-fertilization and complementary learning between the two strategies'.

While knowledge exploitation 'means enhancing the intellectual capital of a company with existing knowledge', knowledge exploration 'is a strategy for a company to increase its intellectual capital by creating its unique private knowledge within its organizational boundary' and therefore 'means enrichment of the intellectual capital that a company achieves by itself' (Ichijo, 2002: 478–9). According to Ichijo (2002), both knowledge exploitation and knowledge exploration are indispensable for a company to increase its competitive advantage and Kyriakopoulos and Moorman (2004: 234) found that – despite the common assumption that these are competing strategies – 'market-oriented firms can gain important bottom-line benefits from pursuing high levels of both strategies in product development'. In fact, '[i]n a world where other firms are seeking to expand their market share, successful firms often can only stay ahead of the competition by exploiting new knowledge to offer improved products or processes that deliver new forms of added value to their customers' (Chaston, 2004: 155). If we interpret Chaston's expression 'exploiting new knowledge' to be a mix of exploiting old knowledge and exploring new knowledge, we might well conclude that his statement is consistent with the above. Vicari and Cillo (2006: 195) follow Kyriakopoulos and Moorman (2004) and define market knowledge exploitation strategies as 'those that imply a leverage on existing knowledge to refine marketing strategies, without exiting the existent path'. On the other hand, they define market exploration strategies as 'those that enact new approaches in the relationship with the market, by challenging existent convictions and routines of the organization' (Vicari and Cillo, 2006: 195–6). Finally, Reinmoeller and van Baardwijk (2005: 63) contend that resilient companies 'go beyond conventional knowledge management by simultaneously exploiting existing knowledge and searching for new knowledge'.

Kyriakopoulos and Moorman (2004: 221) define 'marketing exploitation strategies' as 'strategies that primarily involve improving and refining current skills and procedures associated with existing current skills and procedures associated with existing marketing strategies, including current market segments, positioning, distribution, and other marketing mix strategies' and 'marketing exploration strategies' as 'strategies that primarily involve challenging prior approaches to interfacing with the market, such as a new segmentation, new positioning, new products, new channels, and other marketing mix strategies'. Exploitation strategies have also been referred to as 'adaptive learning' (Senge, 1990; Slater and Narver, 1995) or 'single-loop learning' (Argyris, 1977), and exploration strategies as 'generative learning' (Senge, 1990; Slater and Narver, 1995) or 'double-loop learning' (Argyris, 1977) (cf. also above, 4.1.1.2).

Dynamic capabilities enable 'both the exploitation of existing internal and external firm-specific capabilities and developing new ones' (Teece, Pisano, and Shuen, 1997: 515; cf. also Eisenhardt and Martin, 2000). Kyriakopoulos and Moorman (2004: 222) – who build their concepts of marketing exploitation and exploration on the resource-based view of the firm (cf. also 3.2) – view a firm's market orientation as 'a dynamic capability that facilitates a firm's ability to explore and exploit knowledge and skills'. Indeed, dynamic capabilities are rooted in both exploitative and exploratory activities (Benner and Tushman, 2003). Kyriakopoulos and Moorman (2004: 235, 236) finally conclude that 'as a dynamic capability to sense market changes and relate to markets, a firm's market orientation helps it reconfigure and integrate knowledge generated from both strategies to serve existing and future customer needs' and that 'market orientation is one important firm-level factor that allows high levels of both marketing exploitation strategies (improving current knowledge and skills) and marketing exploration strategies (developing new knowledge and skills) to be used profitably by firms'.

Based on the above, I propose the following definition of 'knowledge-based marketing':

> *Knowledge-based marketing is a knowledge management approach to marketing that focuses both on the exploitation (sharing and application) and exploration (creation) as well as the co-creation of marketing knowledge from contexts, relations, and interactions in order to gain and sustain competitive advantage.*

Note that – even though the term is not mentioned in the definition – the (co-) creation of value is an essential prerequisite for gaining and sustaining competitive advantage. Often, the (co-) creation of knowledge goes hand in hand with the (co-) creation of value, but sometimes it can also be an antecedent. The importance of creating and delivering value has already been emphasized in the definition of marketing given in Chapter 4.2 above and will be further illustrated in the case studies and their discussion in Chapters 5 and 6.

Finally, given the early status of theory-building in knowledge-based marketing, I emphasize that the above is still a preliminary working definition. Significantly I attach equal weight to the role of explicit and tacit marketing knowledge, which are in fact inseparable in marketing practice. Nevertheless, I draw specific attention to the fact that I am giving particular prominence to the role of tacit knowledge (cf. also Chapters 3 and 4.10), in the sense that it has too often been neglected in the past. Besides, following the definition of marketing knowledge above, I also stress that knowledge-based marketing involves stakeholders such as customers, competitors, suppliers, partners, and so on, and is influenced by certain factors, such as national and corporate culture, tacitness of knowledge, and the level of trust (cf. Figure 4.6). Partners include both alliance partners and channel partners. I have attempted with the model below to capture the key features of the processes that I have described in this book.

As shown in Figure 4.6, there are at least four main factors that influence the process of joint exploration and exploitation, as well as

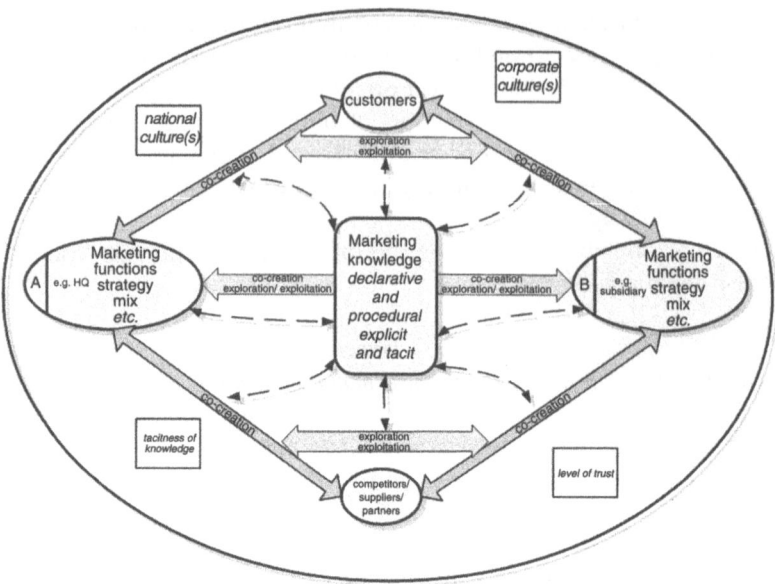

Figure 4.6   Knowledge-based marketing (author's own illustration)

inter- and intra-organizational co-creation of knowledge: national culture, corporate culture, tacitness of knowledge, and level of trust. All of these factors have basically already been discussed in Chapter 3, and partly also Chapter 4. Specifically, Holden's (2002) work on cross-cultural management serves as a helpful framework for analysing the culture-related factors. As far as tacitness is concerned, Simonin (1999a: 469) has noted that 'tacitness is expected to be a strong antecedent of knowledge ambiguity in the process of transferring marketing know-how between partners' and found strong empirical evidence in support of this claim. Cavusgil, Calantone, and Zhao (2003: 9) argue that the 'higher the degree of tacitness of firm knowledge, the harder it is to be transferred from one firm to another' (cf. also Vicari and Cillo, 2006).

It is important to note here that I have not listed 'language' as a separate factor in the model of Figure 4.6, but consider it to be embedded in both national and corporate culture. By corporate culture I mean what Deshpandé and Webster (1989: 4) have termed 'organizational culture' and defined as 'the patterns of shared values and beliefs that help individuals understand organizational functioning and that provide norms for behavior in the organization'. According to Bertels and Savage (1999: 210), 'values and cultural characteristics can foster knowledge creation in terms of product development and customer skills, abilities to learn from customers can enhance relationships, and the ability to learn how knowledge is shared and insights are gained can help to improve communication' (cf. also Davenport and Prusak, 2000). As for trust, it is necessary to recognize that the 'majority of relationships thrive on tacit understanding between parties and only a minority are regulated in contracts' and that commercial relationships are 'usually more informal than formal' and 'the parties trust each other' (Gummesson, 2002: 25). I will come back to some of these influencing factors when discussing relationship marketing in the next section, 4.2.2.2).

Finally, it is also important to note the fact that I have derived the definition and model of knowledge-based marketing from studies of marketing interactions that took place in complex (cross-cultural) contexts (cf. Chapters 5 and 6.1). All too often international marketing is seen as an extension of monocultural marketing (that is, marketing in the firm's domestic market). For far too long the logic has been to apply monoculturally derived concepts of marketing to other countries (Kohlbacher, Holden, Glisby, and Numic, 2007).

4.2.2.2   Key players and relationships

Relationships between customers and suppliers are the ground for all marketing. (Gummesson, 2002: 10)

According to Vicari and Cillo (2006: 185), studies on market orientation and market knowledge are considered 'to address the issue of how companies learn about customers, competitors and channel members in order to continuously sense and act on events and trends in present and prospective markets'. As has become clear from the definition of marketing knowledge (Figure 4.5) and the model of knowledge-based marketing (Figure 4.6), there are at least the following key players and actors involved in the exploration and exploitation (co-creation and sharing) of (marketing) knowledge: different units or subsidiaries of the firm, customers, suppliers, business partners and competitors. Intra-firm knowledge creation and transfer as well as inter-organizational knowledge creation and sharing have frequently been researched and discussed, and I have also reviewed and evaluated the relevant literature in Chapter 3.

In their article 'The relational view' Dyer and Singh (1998) offer a view that suggests that a firm's critical resources may span firm boundaries and may be embedded in inter-firm resources and routines. They argue that an increasingly important unit of analysis for understanding competitive advantage is the relationship between firms, and they identify four potential sources of inter-organizational competitive advantage: (1) relation-specific assets; (2) knowledge-sharing routines; (3) complementary resources/capabilities; and (4) effective governance (Dyer and Singh, 1998). Indeed, relational concepts received more and more research attention in recent years (for example, Brodie, 2002; Chaston, Badger, Mangles, and Sadler-Smith, 2003; Coviello, Brodie, Danaher, and Johnston, 2002; Day, 2000, 2003; Fournier, Dobscha, and Mick, 1998; Glazer, 1991; Griffin and Hauser, 1993; Gulati and Kletter, 2005; Parvatiyar and Sheth, 2000; Peppers and Rogers, 1999; Peppers, Rogers, and Dorf, 1999; Pine II, Peppers, and Rogers, 1995; Sawhney and Zabin, 2002). In an era of globally networked economy, partnership equity is a fundamental component of the relationship equity that a company possesses (Sawhney and Zabin, 2002) and the global partnership base should include the customers, competitors, and suppliers (Yeniyurt, Cavusgil, and Hult, 2005). As Rindova and Fombrun (1999) suggest, the construction of competitive advantage is contingent on both the micro-efforts of the firm, the macro conditions of the environment, and the nature of the firm-constituent interactions (cf. also Tzokas and Saren, 2004). Competitive advantage is therefore built on relationships and relationships with constituents 'are not just exchanges but sustained social interactions in which past impressions affect future behaviors' (Rindova and Fombrun, 1999: 706). In fact, embeddedness in local networks enables companies to gain access to distinct inimitable resources (Eriksson and Chetty, 2003; Schmid and Schurig, 2003; Yeniyurt, Cavusgil, and Hult, 2005). As a result, new organization forms, including strategic partnerships and networks, are replacing simple market-based transactions and traditional bureaucratic hierarchical organizations (Webster, 1992). Indeed, in the industrial age, marketers relied on the framework of the four Ps – product, price, place, and promotion – to develop a marketing plan for their customers and companies created the products and defined their features and benefits; they also set prices, selected places to sell products and services, and promoted intrusively through advertising, public relations, and direct mail. The underlying paradigm was one of uni-directional control (Kotler, Jain, and Maesincee, 2002: 125). The historical marketing management function, based on the microeconomic maximization paradigm, must be critically examined for its relevance to marketing theory and practice and a new conception of marketing will focus on 'managing strategic partnerships and positioning the firm between vendors and customers in the value chain' with the aim of delivering superior value to customers (Webster, 1992: 1). As a result, customer relationships will be seen as the key strategic resource of the business.

Gummesson (2002, 2004b) contends that relationship marketing can offer the beginnings of a general theory, using relationships, networks, and interaction as its core variable, and that there is a need for relationships with customers, suppliers, intermediaries, and competitors. He even makes a call for 'total relationship marketing' and proposes a multilevel approach to relationship marketing that adds theoretical context to relationships, networks, and interaction (Gummesson, 2002). As Kotler, Jain, and Maesincee (2002: 27, original emphasis) put it: 'Companies therefore go beyond the business concept of *customer relationship management* toward the concept of *whole relationship management*. Marketers constantly renew the market by building and managing a customer database and delivering value, with the help of collaborators linked together in a value network.' Gummesson's (2002: 3) definition of relationship marketing is 'marketing based on interaction within networks of relationships'. Data warehousing has emerged from new IT as the next generation of databases, and knowledge residing in organizations can now be collected, stored, and integrated in more elaborate ways. For CRM, this is marketing knowledge about customers; in a broadened relationship marketing sense it is data about all actors in a firm's network (Gummesson, 2001: 33).

For Gummesson (2002: 3), relationship marketing is the 'broader, overriding concept' compared to CRM, which he defines as 'the values and strategies of relationship marketing – with particular emphasis on customer relationships – turned into practical application' (cf. also Day, 2000, 2003; Fournier, Dobscha, and Mick, 1998; Parvatiyar and Sheth, 2000). Relationship marketing is 'grounded in the idea of establishing a learning relationship with each customer' (Peppers, Rogers, and Dorf, 1999: 151; cf. also Peppers and Rogers, 1997; Pine II, Peppers, and Rogers, 1995). This learning relationship is 'an ongoing connection that becomes smarter as the two interact with each other, collaborating to meet the consumer's needs over time' (Pine II, Peppers, and Rogers, 1995: 103). While the basic relationship of marketing is that between a supplier and a customer, a network is 'a set of relationships which can grow into enormously complex patterns' (Gummesson, 2002: 3–4). Gummesson (2002) proposes a total of thirty relationships, among which one is 'the knowledge relationship', which sees knowledge as the most strategic and critical resource.

Tzokas and Saren (2004: 130) suggest that 'the scope, processes and technologies of [relationship marketing] facilitate the process of knowledge construction, embodiment, dissemination and use' and propose a conceptual framework, which they call the 'house of knowledge in relationship marketing' (cf. Figure 4.7). In this framework, 'the knowledge produced by means of interaction and dialogues feeds back to the participants thus giving rise to a new cycle of knowledge creation, dissemination and use' (Tzokas and Saren, 2004: 132). Note the congruence of the influencing factors in this framework and the model of knowledge-based marketing I proposed above (Figure 4.6). Even though I do not explicitly mention commitment it is implicitly incorporated in both trust and corporate culture. Besides, relationship culture and climate are both obviously part of corporate culture (and maybe partly also influenced by national culture).

The process of knowledge creation will be discussed in Chapter 4.2.3 and Figures 4.10, 4.11 and 4.12, as well as Figures 4.13 and 4.14, illustrate the process in detail and highlight the parties involved. These models also fit neatly with Tzokas and Saren's (2004) framework.

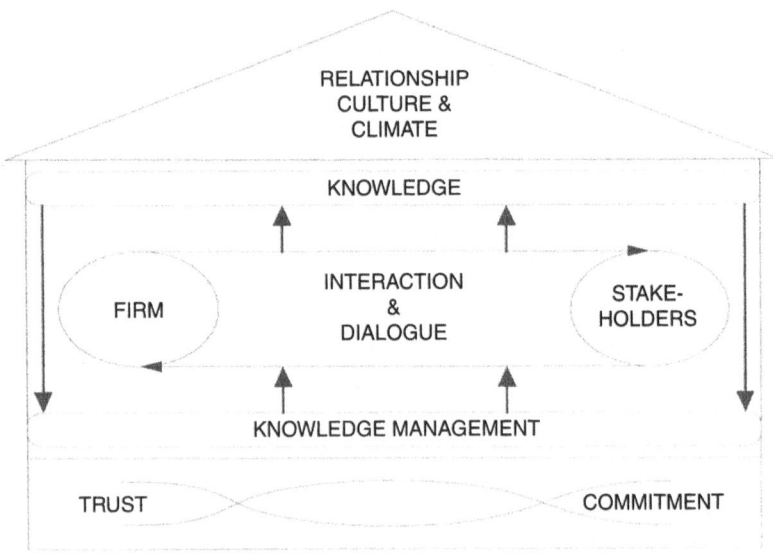

Figure 4.7 The house of knowledge in relationship marketing (from Tzokas and Saren, 2004: 131)

In *Total Relationship Marketing* (Gummesson, 2002) the focal point was relationships, even if networks, the network organization, and the network society played a part. In 'From one-to-one to many-to-many marketing' (Swedish version only), the independent extension of the previous book's focus has moved to the larger context of networks of relationships (Gummesson, 2004a). Gummesson (2004a: 2, removed emphasis) contrasts the two-party relationship, the dyad, with the network and its multiple relationships and defines that many-to-many marketing 'describes, analyses and utilizes the network properties of marketing'. Indeed, the 'array of relationships in the set has been expanded from the dyad of seller and customer to include partners up and down the value chain (e.g., suppliers, the customers of customers, channel intermediaries)' (Day and Montgomery, 1999: 4). Figure 4.8 compares one-to-one marketing (Peppers and Rogers, 1999; Peppers, Rogers, and Dorf, 1999) and many-to-many marketing. The major difference is that the object for one-to-one is a single supplier and single customer relationships, but many-to-many is supplier networks connected with customer networks. The contribution from one-to-one, not least through the expressive wording, is first and foremost to highlight individual interaction in marketing. Indeed, according to Peppers,

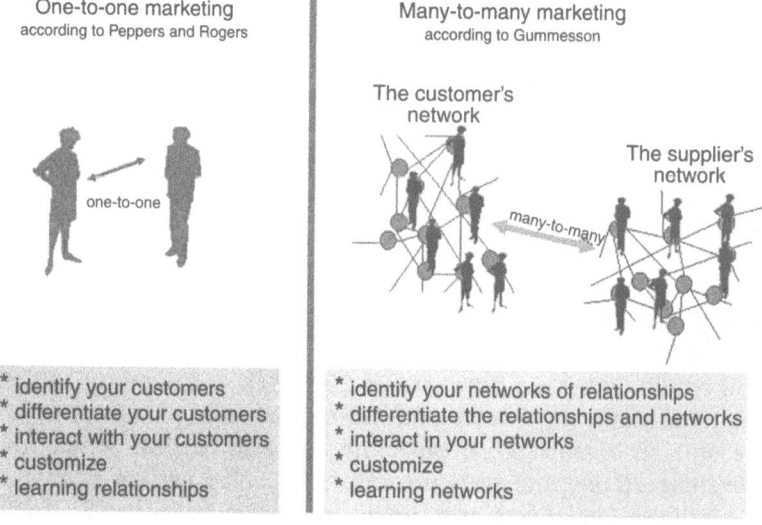

Figure 4.8 Comparison of one-to-one and many-to-many marketing (Gummesson, 2004a: 3; see also Peppers and Rogers, 1999)

Rogers, and Dorf, (1999: 151) one-to-one marketing 'means being willing and able to change your behavior toward an individual customer based on what the customer tells you and what else you know about that customer'. The contribution of many-to-many is taking one-to-one further and addressing the whole context of a complex world (Gummesson, 2004a: 3). Gummesson (2004a: 4, removed emphasis) finally even concludes that many-to-many 'is the closest [he has] come to a DNA of marketing'. Indeed, according to him, marketing 'is interaction in networks of commercial relationships' (Gummesson, 2003b: 168).

Finally, a unique aspect of relationship marketing lies in the fact that it acknowledges the significant role of the customer in the value creation process (Tzokas and Saren, 2004: 129). This has appeared in the literature as value co-production, or prosumer (for example, Rindova and Fombrun, 1999; Wikström, 1996a, 1996b). Value and knowledge co-creation will be discussed further in Chapter 4.2.3.

Gibbert, Leibold, and Probst (2002: 464) state that it is ironic that 'the conceptual predecessor of knowledge management has surpassed its own offspring'. Indeed, ten years ago, proponents of the resource-based view (see also Chapters 3.3 and 4.1.1) of strategy proclaimed that a company is best conceptualized as a bundle of unique resources, or competencies, rather than as a bundle of product market positions (Barney, 1991). More recent contributions to the resource-based view question this one-sided thinking about the locus of competence (Inkpen, 1996; Prahalad and Ramaswamy, 2000). It has now been claimed that such competence has actually moved beyond corporate boundaries, and that it is therefore worthwhile to also look for competence in the heads of customers, rather than only in the heads of employees (Gibbert, Leibold, and Probst, 2002). Therefore, successful companies today work 'with a large set of business partners that make up the company's collaborative network' (Kotler, Jain, and Maesincee, 2002: 118). In fact, 'competence now is a function of the collective knowledge available to the whole system – an enhanced network of traditional suppliers, manufacturers, partners, investors, and customers' (Prahalad and Ramaswamy, 2000: 81). As a result, the organizational ability to develop and nurture interfirm relationships can become an organizational capability and lead to clear competitive advantages (Lorenzoni and Lipparini, 1999; Tzokas and Saren, 2004).

Obviously, nowadays, companies can hardly be viewed as single, independent, and isolated beings any more, and business networks have become ubiquitous in our economy (cf., for example, Iansiti and

Levien, 2004a). Indeed, 'during the last decades of the twentieth century significant changes in our legal, managerial, and technological capabilities made it much easier for companies to collaborate and distribute operations over many organizations' and this development 'pushed many of our industries toward a fully networked structure, in which even the simplest product or service is now the result of collaboration among many different organizations' (Iansiti and Levien, 2004a: 5–6). Consequently, 'large, distributed business networks became *the* established way of doing business in the modern economy' (Iansiti and Levien, 2004a: 6, original emphasis). As a result, increasingly, 'in the new economy, competition is not between companies but rather between collaborative networks, with the prize going to the company that has built the better networks' (Kotler, Jain, and Maesincee, 2002: 24). Indeed, 'the unit of strategic analysis has moved from the single company, to a family of businesses, and finally to what people call the "extended enterprise," which consists of a central firm supported by a constellation of suppliers' (Prahalad and Ramaswamy, 2000: 81). 'The company's position in its network of relationships to customers and lots of other stakeholders – own employees, own suppliers, intermediaries, competitors, allied partners, governments, investors, the media, and others – influences the actual marketing of its products and services' (Gummesson, 2003a: 483). Finally, to 'create new markets, companies may need to draw on resources from collaborators', and rather than 'doing too much on their own, companies will build collaborative networks' (Kotler, Jain, and Maesincee, 2002: 50).

These – more or less – 'loose networks – of suppliers, distributors, outsourcing firms, makers of related products or services, technology providers, and a host of other organizations – affect, and are affected by, the creation and delivery of a company's own offerings' (Iansiti and Levien, 2004b: 69). As Chaston (2004: 21) puts it: 'in the twenty-first century, it can confidently be predicted that knowledge networks of various forms will become an increasingly dominant operational structure through which to ensure the effective management of entrepreneurial activities in both private and public sector organisations'. Indeed, companies and markets 'are networks of relationships within which we interact, completely in accordance with the definition of [relationship marketing]' (Gummesson, 2002: 8). Given this situation, a company's success depends on the success of its partners (Iansiti and Levien, 2004a). In fact, an 'active partnership between companies and their customers, collaborators, and communities will help companies maximize company-delivered value and reduce company-delivered costs, as well as help companies respond faster to emerging opportunities' (Kotler, Jain, and Maesincee, 2002: 52). Moreover, 'neither value nor innovation can any longer be successfully and sustainably generated through a company-centric, product-and-service-focused prism' (Prahalad and Ramaswamy, 2003: 12).

It is therefore not surprising that Nonaka and Toyama (2003: 8) argue that *ba* is 'not limited to the frame of a single organization but can be created across the organizational boundary', for example, as a joint venture with a supplier, an alliance with a competitor, or an interactive relationship with customers, universities, local communities, or the government (cf. also Kokuryo, Nonaka, and Kataoka, 2003). A firm has therefore been identified as the organic configuration of *ba* (Nonaka, Toyama, and Nagata, 2000: 8–9; Nonaka and Toyama, 2002: 1001, 1006) or an 'organic configuration of multilayered *ba*' (Nonaka and Toyama, 2005: 429) (cf. Figure 4.9).

Relevant marketing information can arise from a variety of external sources (Barabba and Zaltman, 1991; Kohli and Jaworski, 1990; Moorman, 1995) and a business must be careful not to underestimate the potential contributions of other learning sources, such as suppliers, businesses in different industries, consultants, universities, government agencies, and others that possess knowledge valuable to the business

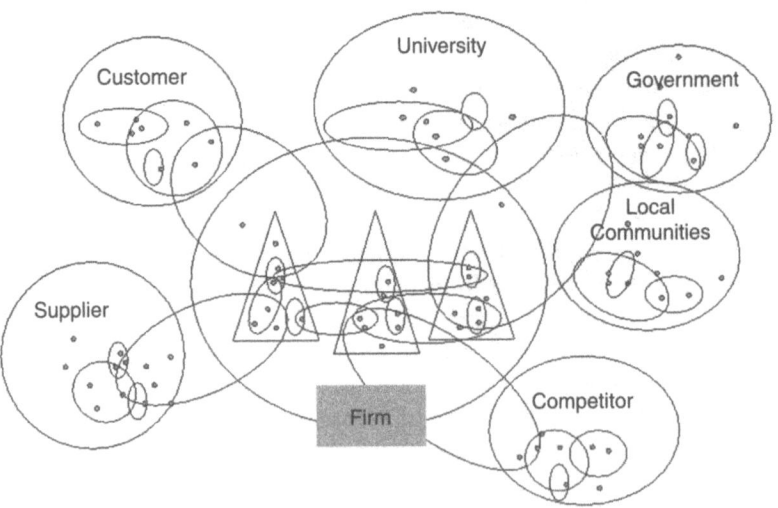

Figure 4.9 Organization as organic configuration of *ba* (from Nonaka and Toyama, 2003: 8)

(Achrol, 1991; Desouza, Awazu, and Jasimuddin, 2005; Dickson, 1992; Kanter, 1989; Slater and Narver, 1995; Webster, 1992). Desouza, Awazu, and Jasimuddin (2005: 16) put it like this: 'Most organizations need to concern themselves with external sources of knowledge from suppliers, business partners, customers, government and regulatory bodies, academia and competitors.' Business partners, for example, like suppliers, 'have deep knowledge in their areas of focus, as this represents their bread and butter'; academia also 'represents a viable external source of knowledge for business organizations' and organizations 'must also get knowledge from competitors' (Desouza, Awazu, and Jasimuddin, 2005: 17, 18). Resources are increasingly being built through networks of co-operating companies and collaboration becomes a key marketing strategy (Gummesson, 2002). Besides, external knowledge sources 'can come from highly explicit to highly tacit sources but either way, gathering knowledge from new ideas or emerging innovations normally calls for discussions, which should be face to face' (Desouza, Awazu, and Jasimuddin, 2005: 19). Thus, learning, teaching, and transferring knowledge across boundaries will become essential skills (Prahalad and Ramaswamy, 2000: 81).

According to Gummesson (2004a: 9), organizations are dependent on alliances, outsourcing, knowledge-sharing, and so on, to be competitive, which requires a network approach to organization. These networks have also been termed (business) 'ecosystems' (cf., for example, Iansiti and Levien, 2004a, 2004b; Nonaka and Toyama, 2005). As a result, the 'economic value of a knowledge-creating firm is created through the interactions among knowledge workers, or between knowledge workers and the environment such as customers, suppliers, or research institutes' (Nonaka and Toyama, 2005: 430) and the 'firm's knowledge base includes its technological competences as well as its knowledge of customer needs and supplier capabilities' (Teece, 1998: 75, 2000b: 38). These networks can frequently be similar to the small world networks described by Watts (2003). Doz, Santos, and Williamson (2003: 163) speak of a 'sensing network' 'that could identify innovative technologies or emerging customer needs', in short 'a network that pre-empted global sources of new knowledge' (cf. also Doz, Santos, and Williamson, 2001). But it is important to note that 'sensing' 'is much more than market research or information gathering' as 'it involves accessing complex knowledge that is often tacit and deeply embedded in a local context' (Doz, Santos, and Williamson, 2003: 163). Besides, it is 'clearly not enough for a company to amass a rich hoard of knowledge from around the world' and only to access dispersed knowledge, but the metanational has to 'also mobilize it to create innovative products, services, processes, and business models' (Doz, Santos, and Williamson, 2003: 164). This requires building a set of structures to translate new knowledge into innovative products or specific market opportunities, and these structures within the network are termed 'magnets' (Doz, Santos, and Williamson, 2001, 2003). These magnets may take various organizational forms and 'attract dispersed, potentially relevant knowledge and use it to create innovative products, services, or processes, and they then facilitate the transfer of these innovations into the network of day-to-day operations' (Doz, Santos, and Williamson, 2003: 165). As a result, 'competence now is a function of the collective knowledge available to the whole system – an enhanced network of traditional suppliers, manufacturers, partners, investors, *and* customers' (Prahalad and Ramaswamy, 2000: 81, original emphasis; cf. also Prahalad and Ramaswamy, 2003, 2004a). Indeed, external knowledge is 'more and more important than ever for organizations to be sufficiently competitive in the current and future market' (Desouza, Awazu, and Jasimuddin, 2005: 19). Obviously, the concept of CoPs (cf. 3.5) – specifically those with members across different organizations – will also play an important role here.

According to Nonaka and Toyama (2005: 430, original emphasis), the 'ecosystem of knowledge consists of multi-layered *ba*, which exists across organizational boundaries and is continuously evolving', with firms creating knowledge 'by synthesizing their own knowledge and the knowledge embedded in various outside players, such as customers, suppliers, competitors or universities'. Therefore, 'establishing a carefully planned network of alliances with lead customers, suppliers, universities, research institutes and even competitors in various parts of the world can be an invaluable aid in prospecting for new market knowledge or technical know-how' (Santos, Doz, and Williamson, 2004: 35).[29] Through interactions with the ecosystem, a firm creates knowledge, and the knowledge created changes the ecosystem (Nonaka and Toyama, 2005: 430). Indeed, '[v]aluable knowledge can often come from the periphery of an organization, where very different environments tend to encourage diverse skills and capabilities' (Santos, Doz, and Williamson, 2004: 35; cf. also Prahalad and Hamel, 1990; Stalk, Evans, and Shulman, 1992; as well as the concept of peripheral vision: Day and Schoemaker, 2006; Long Range Planning, 2004). Therefore, 'in the future the competitive advantage of the multinational enterprise will come, not so much from its efficiency in transferring resources, information and knowledge, but from its unique potential for radical innovation by melding and leveraging distinctive knowledge drawn from diverse geographical contexts around the world' (Doz, Santos, and Williamson, 2003: 155). Therefore, truly global knowledge-based companies have to become 'metanational' (Doz, Santos, and Williamson, 2001, 2003; Santos, Doz, and Williamson, 2004). As a result, the 'new frontier for managers is to create the future by harnessing competence in an enhanced network that includes customers' (Prahalad and Ramaswamy, 2000: 87; cf. also Prahalad and Ramaswamy, 2003, 2004a).

In a similar vein, Chaston (2004: 17) proposes a 'hub structure' or 'hub knowledge network' as an approach to building knowledge networks. The role of the central organization is to bring together knowledge exchange between market system members such as suppliers, intermediaries, and customers. Indeed, 'relationships help create unique, difficult to imitate knowledge for firms' (Tzokas and Saren, 2004: 125).

Finally, learning from others encompasses common practices, such as benchmarking, forming joint ventures, networking, making strategic alliances, and working with lead customers, who both recognize strong needs before the rest of the market and are motivated to find solutions to those needs (for example, Kanter, 1989; Slater and Narver, 1995; Webster, 1992). Because learning organizations have close and extensive relationships with customers, suppliers, and other key constituencies, there is a co-operative attitude that facilitates mutual adjustment among them when the unexpected occurs (Slater and Narver, 1995; Webster, 1992).

4.2.3 Marketing knowledge co-creation

'Collaboration has become an established way of doing business with suppliers, channel partners and complementors', but, with a few exceptions, 'working directly with *customers* to co-create value remains a radical notion' (Sawhney, 2002: 96, original emphasis). But a 'critical aspect of creating a successful market is the ability to integrate the customer into every key process' and collaborators 'may play a major role in initiating knowledge creation in the marketspace' (Kotler, Jain, and Maesincee, 2002: 36, 38). According to Achrol and Kotler (1999), the creation of marketing know-how is the most important function of marketing in the global knowledge-based economy. Indeed, 'in marketing, a wide array of knowledge needs to be created' and 'knowledge on customers and their preferences must be located or solutions for a particular kind of customer problem need to be identified' (Schlegelmilch and Penz, 2002: 12). In the four sections of Chapter 4.1.3, I have looked at knowledge creation and application processes in different points of the value chain. But most of the time the knowledge creation or application is only conducted in a unilateral, one-sided way. Firms generate, collect, and analyse knowledge about customers, customers' needs, competitors, suppliers, and so on. Customer knowledge from customers can be seen as a small exception to this, but here as well, the knowledge might be communicated unilaterally from the customers to the firm. But the real challenge and source of essential knowledge for competitive advantage might be to go beyond knowledge creation and application as a unilateral concept. In fact, interactions and knowledge co-creation might become more and more crucial. To quote Gummesson (2002: 26): 'In new marketing and management theory, the relationship is increasingly seen as interaction and joint value creation. The content of a relationship is often knowledge and information.'

Therefore, knowledge and value co-creation with customers – but also with suppliers and other business partners – has also received significant attention recently (cf., for example, Doz, Santos, and Williamson, 2001, 2003; Gummesson, 2002; Lawer, 2005; Prahalad and Ramaswamy, 2000, 2003, 2004a; Sawhney, 2002; Sawhney and Prandelli, 2000a; Thomke and von Hippel, 2002; Wikström, 1996a, 1996b; Zack, 2003). Indeed, according to Prahalad and Ramaswamy (2000: 80), the market has become 'a forum in which consumers play an active role in creating and competing for value', with the distinguishing feature of this new marketplace being 'that consumers become a new source of competence for the corporation' (cf. also Prahalad and Ramaswamy, 2003, 2004a). In fact, 'co-creation converts the market into a *forum* where dialogue among the consumer, the firm, consumer communities, and networks of firms can take place' (Prahalad and Ramaswamy, 2004a: 122, original emphasis). As a result, we must 'view the market as a *space of potential co-creation experiences* in which individual constraints and choices define their willingness to pay for experiences', that is, 'the market resembles a forum for co-creation experiences' (ibid., original emphasis). According to Zack (2003: 71), anyone who can help the business – customers, trading partners, suppliers, consumers, interest groups – should be involved to create the knowledge the company needs. Indeed, as discussed above, the 'array of relationships in the set has been expanded from the dyad of seller and customer to include partners up and down the value chain (e.g., suppliers, the customers of customers, channel intermediaries)' (Day and Montgomery, 1999: 4). Figure 4.10 illustrates

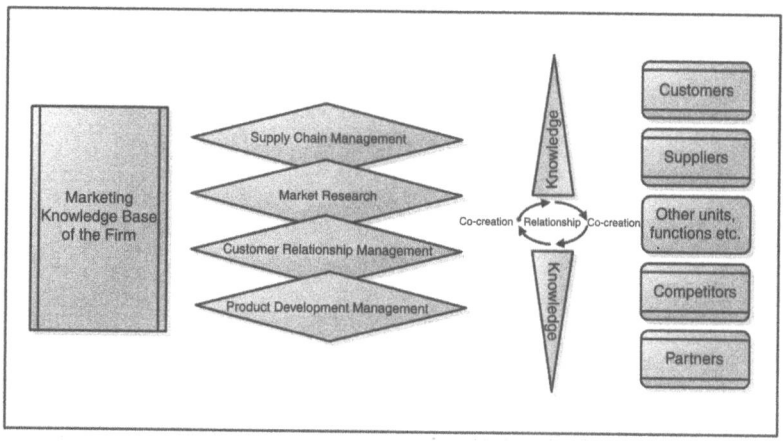

Figure 4.10   Knowledge-based marketing processes (author's own illustration)

these relationships and the knowledge-based marketing processes along the value chain. Note that the marketing knowledge box is meant to represent the marketing knowledge – or marketing knowledge base – of one particular firm in general. Partners, as always in this book, include both alliance partners and channel partners. Finally, also the strategy-making process 'should incorporate diverse inputs, including insights about and from customers, competitive information, views of outside experts, and fresh thinking about new technologies that might disrupt the business' (Day and Schoemaker, 2006: 145).

As proposed in Chapter 4.1.3, SCM, market research, CRM, and product development are interdependent and interwoven processes. They mutually benefit from each other's knowledge and should be managed in an integrated and comprehensive way. Figure 4.11 therefore summarizes them as marketing processes in general. A further generalization and simplification is achieved by grouping competitors, suppliers, and partners together. Basically, these three, together with the customers, are all stakeholders of the company. But for obvious reasons, the special position and meaning of customers for the company is highlighted. Figure 4.12 illustrates this in consistence with Figure 4.6

Gibbert, Leibold, and Probst (2002: 463) contend that since CKM is about innovation and growth, customer knowledge managers 'seek opportunities for partnering with their customers as equal co-creators of organizational value'. According to Prahalad and Ramaswamy (2003), the value of products or services is in the co-creation experience

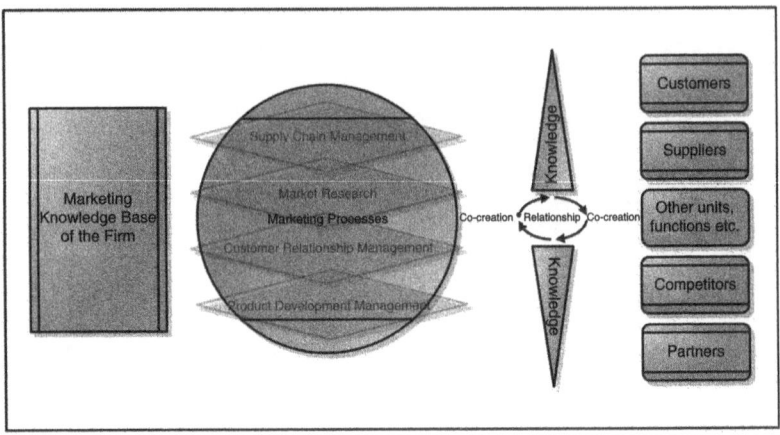

Figure 4.11 Knowledge-based marketing processes (integrated model I) (author's own illustration)

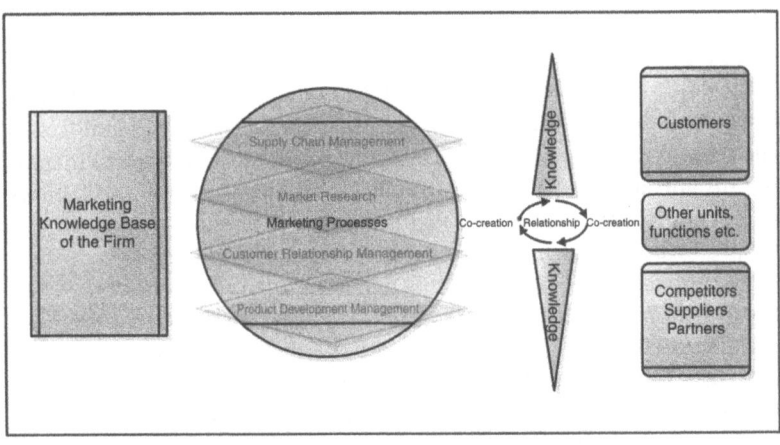

Figure 4.12 Knowledge-based marketing processes (integrated model II) (author's own illustration)

that stems from the customer's interaction with the product and/or the firm (cf. also Prahalad and Ramaswamy, 2004a). Gummesson (2002: 8) further notes that '[e]specially in services and often in B-to-B, customers are co-producers'. Lovelock and Gummesson (2004: 29) use the term 'coproducer' in the narrow sense of 'a transfer of work from the provider to the customer' and contend that '[i]n its purest form, coproduction means that customers engage in self-service, using systems, facilities, or equipment supplied by the service provider'. But research in this area is still rather scarce, and as Lawer (2005: 11) has noted, 'the organizational learning or marketing literature does not yet adequately define or empirically identify the nature or scope of the capability changes required for co-creation of knowledge with customers'. Indeed, 'the challenge is to view customers as co-producers of knowledge' (Desouza and Awazu, 2005b: 143) and in order 'to be successful at co-producing knowledge, the organization must seek customers who have open knowledge-sharing cultures, are willing to engage in learning and knowledge–creating activities, and are willing to take a certain degree of risk' (Desouza and Awazu, 2004: 15). Finally, companies must 'redesign their businesses from a customer-driven starting point, so that they gather deep knowledge about customers and then have the capacity to offer customized products, services, programs, and messages' (Kotler, Jain, and Maesincee, 2002: 164).

Communities of creation as a CKM style (cf. also 4.1.3.3) are reflected by the process of putting together customer groups of expert knowledge that interact not only with the company, but importantly, also with each other (Gibbert, Leibold, and Probst, 2002; Sawhney and Prandelli, 2000b; Wikström, 1996b). Sawhney and Prandelli (2000b) describe the practice of distributed innovation within such communities of creation and suggest that the knowledge required to compete in technology markets is becoming more diverse whilst, at the same time, firms are increasingly narrowing their knowledge base in an effort to specialize and focus. In such an environment, firms can no longer produce knowledge autonomously but rather must co-operate with their trading partners and customers to create knowledge. Such a distributed approach to learning and innovation, they suggest, requires new innovation governance mechanisms, one of which is the communities of creation model – a knowledge socialization mechanism that sits between complete open-source, market-based approaches for innovation and closed, autonomous firm-based approaches (Sawhney and Prandelli, 2000b; cf. also Lawer, 2005). The community of creation model is grounded in the concept of *ba*: 'participating in a ba means transcending one's own limited perspective or boundary and contributing to a dynamic process of knowledge development and sharing. Similarly, participating in a community of creation involves socializing one's individual knowledge and contributing to the creation of a joint output that is superior to the sum of the individual outputs, because new knowledge is created through the emerging relationships' (Sawhney and Prandelli, 2000b: 25). Thus, similar to CoPs (cf. 3.5), communities of creation are groups of people, first who work together over a long period of time, second they have an interest in a common topic, and third, they want to jointly create and share knowledge. Unlike the traditional CoP, however, communities of creation span organizational, rather than functional boundaries to create common knowledge and value (Gibbert, Leibold, and Probst, 2002). Indeed, '[i]n co-creation, direct interactions with consumers and consumer communities are critical' (Prahalad and Ramaswamy, 2004b: 11).

Furthermore, customer interactions (Furukawa, 1999a, 1999b; Vandenbosch and Dawar, 2002) and customer experiences (Berry, Carbone, and Haeckel, 2002; Carbone and Haeckel, 1994; Pine II and Gilmore, 1999; Prahalad and Ramaswamy, 2000, 2003, 2004a) have become key terms in this context. In fact, '[h]igh-quality interactions that enable an individual customer to co-create unique experiences with the company are the key to unlocking new sources of competitive advantage' (Prahalad and Ramaswamy, 2004b: 7). Prahalad and Ramaswamy (2003: 15) propose the concept of 'experience environment', which 'can be thought of as a robust, networked combination of company capabilities ... and consumer interaction channels ... flexible enough to accommodate a wide range of individual context-and-time-specific needs and preferences'. The network creates an experience environment with which each customer has a unique interaction. The consumer actively co-creates his or her personalized experience, which forms the basis of value to that consumer (Prahalad and Ramaswamy, 2003: 15). Because we must continually co-create new knowledge to co-create value continually, so-called 'knowledge environments' for managers resemble experience environments for consumers (Prahalad and Ramaswamy, 2004a: 171). To be effective, 'a knowledge environment must engage the total organization, including multiple levels, functions, and geographies', and the knowledge environment 'is also where *the manager, as consumer, interacts with the experience network* to co-create value' (Prahalad and Ramaswamy, 2004a: 179, 185, original emphasis).

Customers' ideas – specifically those of so-called 'lead users' (for example, Franke, von Hippel, and Schreier, 2006; von Hippel, 1977, 1986, 1988, 1994, 2006) – and the ideas of those that interact directly with customers, or those that develop products for customers, have become important (cf., for example, Barabba and Zaltman, 1991; Leonard, 1998, 2000, 2006; Schrage, 2006; Zaltman, 2003). 'Lead users have foresight (knowledge) to help an organization better plan for product innovations' and organizations have 'begun to host user conferences for the specific purpose of getting to know how their customers utilize their products and how they have customized or modified them to meet their needs' (Desouza and Awazu, 2004: 14). In the 1970s, von Hippel (1977) found that most product innovations came not from within the company that produced the product but from end-users of the product. Note that lead users can be part of or can also form networks and share their ideas and knowledge within them (Furukawa, 1999a, 1999b). More recently, Thomke and von Hippel (2002) suggested ways in which customers can become co-innovators and co-developers of custom products (cf. also Gibbert, Leibold, and Probst, 2002; Thomke, 2003). Indeed, '[c]ontrary to the mythology of marketing, the supplier is not necessarily the active party' and in B-to-B, 'customers initiate innovation and force suppliers to change their products or services' (Gummesson, 2002: 15). As Prahalad and Ramaswamy (2004b: 10–11, original emphasis) put it: 'In the co-creation view, all points of interaction between the company and the consumer are opportunities *for both value creation and extraction.*'

In the traditional conception of the process of value creation, customers were 'outside the firm' and value creation occurred inside the firm (through its activities) and outside markets (Prahalad and Ramaswamy, 2004b: 6; cf. also Prahalad and Ramaswamy, 2000, 2004a). Indeed, Porter's (1980) concept of the value chain 'epitomized the unilateral role of the firm in creating value' (Prahalad and Ramaswamy, 2004b: 6). However, in the knowledge economy, 'companies must escape the firm-centric view of the past and seek to co-create value with customers through an obsessive focus on personalized interactions between the consumer and the company' (Prahalad and Ramaswamy, 2004b: 7; cf. also Savage, 1996). Sheth, Sisodia, and Sharma (2000: 62) propose the concept of co-creation marketing:

> With an increase in customer-centric marketing, customers will have an increasing role in the fulfillment process, leading to 'co-creation marketing'. Cocreation marketing involves both the marketers and the customer who interact in aspects of the design, production, and consumption of the product or service.

Sheth, Sisodia, and Sharma (2000) also suggest that the extent of co-creation marketing depends on how much customer knowledge a company is able to accumulate and use. Co-creation marketing enables and empowers customers to aid in product creation, pricing, distribution, and fulfilment and communication. It can enhance customer loyalty and reduce the cost of doing business (cf. also Lawer, 2005). Last but not least, Vargo and Lusch (2004) develop links between value co-creation and the service-centred model of marketing, which reflects a view of marketing that means more than simply being customer oriented; rather, it means collaborating with and learning from customers by being adaptive to their individual and dynamic needs. They contrast this with the more conventional value creation process where companies and consumers had distinct roles of production and consumption. In this scenario, products and services contained value and markets exchanged this value, from the producer to the consumer. Value creation occurred outside markets. By contrast, the service-centred marketing logic implies that value is defined by and co-created with the customer rather than embedded in physical products (Lawer, 2005; Prahalad and Ramaswamy, 2000, 2003, 2004b; Vargo and Lusch, 2004).

From the above, it should have become clear that a knowledge-based approach to marketing asks for the co-creation of knowledge – and subsequently the co-creation of value – with a variety of key players and actors in the business ecosystem. Zack (2003: 69) puts it like this:

> Knowledge creation and sharing in today's economy are not bound by the traditional physical and legal limits of the corporation. Companies are increasingly realizing that knowledge is often produced and shared as a byproduct of daily interactions with customers, vendors, alliance partners and even competitors. The knowledge-based organization, then, is a collection of people and supporting resources that create and apply knowledge via continued interaction.

As mentioned above, for Achrol and Kotler (1999), the creation of marketing know-how is the most important function of marketing in the global knowledge-based economy. Indeed, '[s]ince knowledge is socially constructed, focus on knowledge creation, rather than knowledge transfer, becomes paramount for organizational learning' (Plaskoff, 2003: 164). Besides, 'in marketing, a wide array of knowledge needs to be created' and 'knowledge on customers and their preferences must be located or solutions for a particular kind of customer problem need to be identified' (Schlegelmilch and Penz, 2002: 12). For the latter task, CRM and data mining tools for decision support have proven useful (Shaw, Subramaniam, Tan, and Welge, 2001; Wierenga and Ophuis, 1997), and as Shaw and fellow researchers (2001) have shown, can be integrated into a marketing knowledge framework. But even though this gathering and systemizing of marketing knowledge can also be seen as a form of knowledge generation – and as such plays an important role for marketing functions and tasks – it should not be mistaken for the innovative process of organizational knowledge creation depicted and analysed by Nonaka and Takeuchi (1995).

By organizational knowledge creation, Nonaka and Takeuchi (1995: 3) mean 'the capability of a company as a whole to create new knowledge, disseminate it throughout the organization, and embody it in products, services, and systems' by organizational knowledge creation and they develop a dynamic model of this process (SECI model). As – in the strict sense – knowledge is created only by individuals, organizational knowledge creation 'should be understood as a process that "organizationally" amplifies the knowledge created by individuals and crystallizes it as a part of the knowledge network of the organization' (Nonaka and Takeuchi, 1995: 59). Referring to Brown and Duguid's (1991) work on 'evolving communities of practice' (cf. also Chapter 3.5), Nonaka and Takeuchi (1995: 59) point to the fact that this process of organizational knowledge creation 'takes place within an expanding "community of interaction", which crosses intra- and inter-organizational levels and boundaries'.

Figure 4.13 illustrates the interconnected processes of organizational knowledge creation and marketing strategy, that is, the marketing knowledge co-creation process. Through the SECI process (cf. 3.4), new knowledge is constantly created and refined over time, lifting the knowledge from the tacit and explicit organizational knowledge base to a higher dimension, namely in the form of holistic knowledge. In a sense, this holistic knowledge bridges explicit and tacit knowledge and can therefore be seen as a kind of synthesis of both. Indeed, bridging the gap between explicit and tacit knowledge means bridging the gap 'between the formula and its enactment'. Taylor (1993: 57) contends that the 'person of real practical wisdom is marked out less by the ability to formulate rules than by knowing how to act in each particular situation'. This is consistent with the definitions of knowledge presented in Chapters 3.1 and 4.2.1. As will be recalled, Dixon (2000: 13) defines knowledge 'as the meaningful links people make in their minds between information and its application in action in a specific setting' and states that it 'is always linked to action', as it is 'derived from action and it carries the potential for others to use it to take action'. In a similar vein, tacit knowledge refers to a kind of knowledge which is highly personal, hard to formalize, and thus difficult to

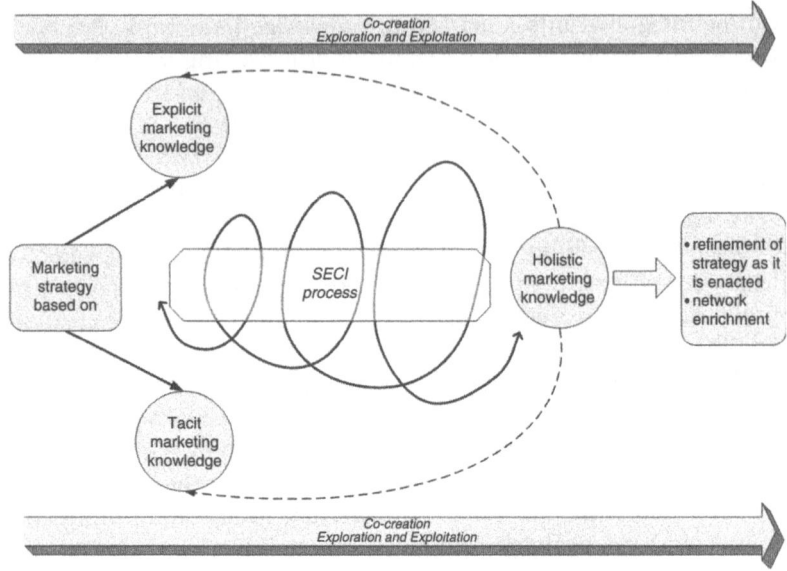

Figure 4.13 The marketing knowledge co-creation process (author's own illustration)

communicate to others, as it is deeply rooted in action (Nonaka, 1996: 21). Indeed, in management, 'knowledge about situations is of prime importance ... not just knowledge about facts or people or technology, et cetera, but situational knowledge that combines all these factors' (Ghosn and Riès, 2005: 175).

Note that the marketing knowledge co-creation process in Figure 4.13 is exactly the knowledge co-creation process (and actually also exploration and exploitation process) that can be found in the models of marketing knowledge (Figure 4.5), knowledge-based marketing (Figure 4.6), and the knowledge-based marketing processes (Figure 4.10). In a sense, the knowledge-based marketing model (Figure 4.6) is a macro model of knowledge-based marketing as proposed in this book. The knowledge-based marketing processes model (Figure 4.10) is a model on the meta level, while the model of marketing knowledge (Figure 4.5) and the model of the marketing knowledge co-creation process (Figure 4.13) are micro models that explain concepts (marketing knowledge) and processes (knowledge co-creation) incorporated in the macro and meta models in greater detail.

Finally, as shown in Chapter 3.4, organizational knowledge creation needs a shared context/*ba* or it is at least enhanced by it. Figure 4.14 takes

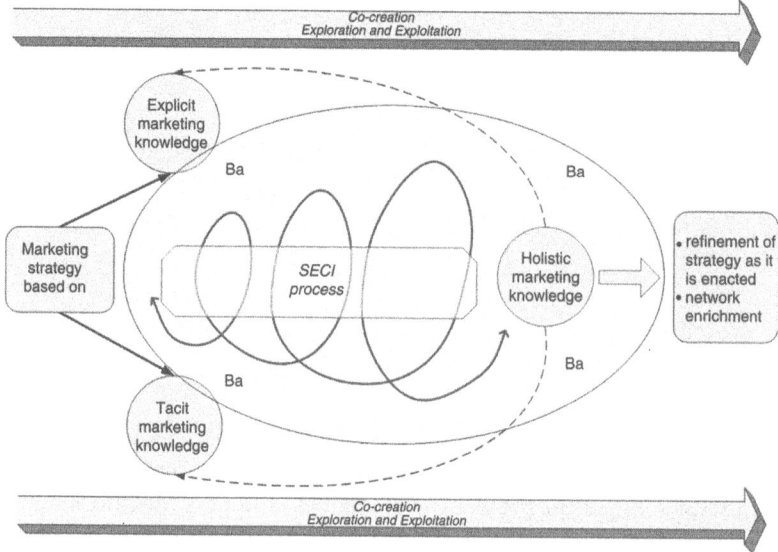

Figure 4.14 The marketing knowledge co-creation process and *ba* (author's own illustration)

this fact into consideration and incorporates *ba* into the model. For the case of the interaction and subsequent knowledge exchange and creation between firms and customers, Furukawa (1999a, 1999b) proposes the concept of 'meeting *ba*' (*deai no ba*). This meeting *ba* needs to be designed to communicate actively with customers and consumers within the social network.

5
Case Studies

This chapter, after a brief overview, presents the case studies which will be discussed and analysed in Chapter 6.1. This chapter seeks to clarify and discuss some of the problems, possibilities, and risks the informant companies face, and analyses six explanatory case studies of knowledge-based approaches to marketing. Research methodology and the empirical research project are described in the Appendix.

This chapter illustrates how things could and can be done differently in two senses. First, it shows how six companies have opted to pursue knowledge-based approaches to marketing and thus act differently from many of their competitors. And second, the approaches of the six companies – despite some similarities – also differ from each other and thus offer insights on different knowledge-based strategies.

5.1 Overview

In the course of conducting research for this book, of 35 companies, nine were selected for in-depth case studies, and of these nine, six were selected to serve as explanatory cases studies of knowledge-based marketing (for details see Appendix A.2.2 on sampling). Figure 5.1 gives a brief overview of the six case studies.

The case studies are presented in the form of abbreviated vignettes (Stake, 1995; Yin, 2003a), illustrating the essence of each informant company's knowledge-based approach. In fact, as they were conducted as explanatory case studies (Yin, 2003a, 2003b), they are meant to highlight how the six informant companies have adopted and implemented especially distinguished knowledge-based approaches to marketing. The vignettes are all built to the following structure:

	Informant company	Case	Location(s)
1	Hewlett-Packard Consulting & Integration	Learning Community (CoP)	Japan, Austria
2	Schindler Elevator	Escalator Product Launch in Asia Schindler 9300 AE Escalator	Japan, Hong Kong, Austria
3	Siemens	Siemens One	Japan, Germany, Austria, China
4	Toyota Peugeot Citroën Automobile (TPCA)	Inter-organizational knowledge creation and learning in an IJV	Czech Republic, Japan
5	Mazda	New product development Mazda Roadster Miata	Japan
6	Maekawa Manufacturing	Knowledge and value co-creation with customers	Japan

Figure 5.1 Overview of the case studies.

- company information/ background;
- knowledge management initiatives/activities in general – brief overview (if applicable);
- knowledge-based marketing case:
  - case background;
  - case study proper: knowledge creation, sharing, transfer etc., and so on;
- conclusion.

Storytelling has become a popular method for sharing and transferring knowledge in organizations (cf., for example, Colton and Ward, 2004; Schreyögg and Geiger, 2006; Swap, Leonard, Shields, and Abrams, 2001, to name but a few). Case studies usually also have a story to tell. But this should not be a lengthy narrative with many – often too many and confusing – details. Using abbreviated vignettes aims at presenting only the essence of the case with a clear focus on the research question. Nevertheless, narrative inquiries 'develop descriptions and interpretations of the phenomenon from the perspective of participants, stakeholders, researchers, and others' (Flyvbjerg, 2006b: 380). Gummesson (2001, 2005) encourages 'narrative research' in marketing. For him, narratives 'are accounts – stories – about experiences, and they can take many forms' and by 'presenting research as a story, we avoid the fragmentation that is inevitable when we break down networks of events into abstract concepts and categories' (Gummesson, 2005: 324). Probst (2002: 318) sees case-writing as a knowledge management and organizational learning tool, arguing that narrative case studies 'put tacit knowledge to work'.

Finally, there are two important points to keep in mind. First, in the fast moving and quickly changing business environment of the know-ledge economy, the case vignettes presented below cannot be but mere snapshots of business activities and situations from the past. I have tried my best to use the most up-to-date information, but even while writing this chapter, the world outside kept changing and moving on. But the most important issue is not to give the most accurate and current description of what is going on – for this we can refer to the Internet, newspapers, TV, and other media, even though they, despite their frequent updates, face a challenge in keeping track – but to give an account of explanatory cases which, by the 'force of example' (Flyvbjerg, 2006a), help to illustrate, at least partially, the ideas and concepts put forward in Chapter 4.2. Second, rather than seeing the case studies below as best practices, I view them simply as explanatory practices, maybe also offering a first glimpse of 'next practices' (Prahalad and Ramaswamy, 2004a) that show how some leading global companies are on their way to successfully implementing and leverage-ing knowledge-based marketing for competitive advantage.

5.2 Hewlett-Packard Consulting & Integration

This case study[30] illustrates how Hewlett-Packard (HP) Consulting & Integration (CI) and specifically HP CI Japan leverage a particular form of CoP (see also 3.5) to create and share both tacit and explicit cus-tomer and other marketing knowledge.

5.2.1 Company information/background

HP consists of four global business groups with 150 000 employees in more than 170 countries, and total revenue of approximately US$ 87 billion in financial year (FY) 2005. HP's corporate activities in Japan go back to 1963 and HP Japan is HP's legal corporate entity in Japan with 5600 employees and a turnover of almost 412 billion yen (approx. US$3.5 billion) as of November 2005. HP Consulting & Integration (CI) is part of HP Services (Technology Solutions Group), which has 65 000 IT professionals in 160 countries around the world encompassing four geographical regions (Americas, Asia Pacific, EMEA, Japan). Its main business is the system integration (SI) of corporate computer systems, which includes the development of system software for customers, IT consulting, sales, and distribution of software developed by HP and other developers.

HP has frequently been featured as a role model in numerous books and articles on knowledge management and has also been a recipient of the MAKE (Most Admired Knowledge Enterprise) award several times (cf., for example, English and Baker, 2006).

5.2.2 Knowledge management initiatives/activities in general

At HP CI, knowledge management is a systematic approach to help information and knowledge flow to the right people at the right time so that they can act more efficiently and effectively in their daily job. The knowledge management programme relies on three main components: people who are the producers and consumers of knowledge, processes that guide the management of the knowledge, and technology/tools to facilitate access to knowledge assets (see Figure 5.2).

HP CI's knowledge management activities can be divided into three different levels. On level 1, the *@hp employee portal* can be accessed by all HP employees worldwide and across all business groups. It is integrated into HP's intranet and used for general communication and information sharing. Level 2 consists of different global repositories and communities. It will be discussed below. On level 3, different collaboration tools and team workspaces for virtual collaboration of teams and team members from different locations can be found.

HP CI's knowledge management activities are managed and controlled by its knowledge management departments and their knowledge managers and knowledge advisers. While the knowledge managers' task is to implement the worldwide strategy and tools through communication and marketing, training and consulting, building interfaces (HR, IT, Marketing, Project Management Office) and reward and recognition programmes, knowledge advisers give assistance on knowledge management processes and tools, direct people to the right

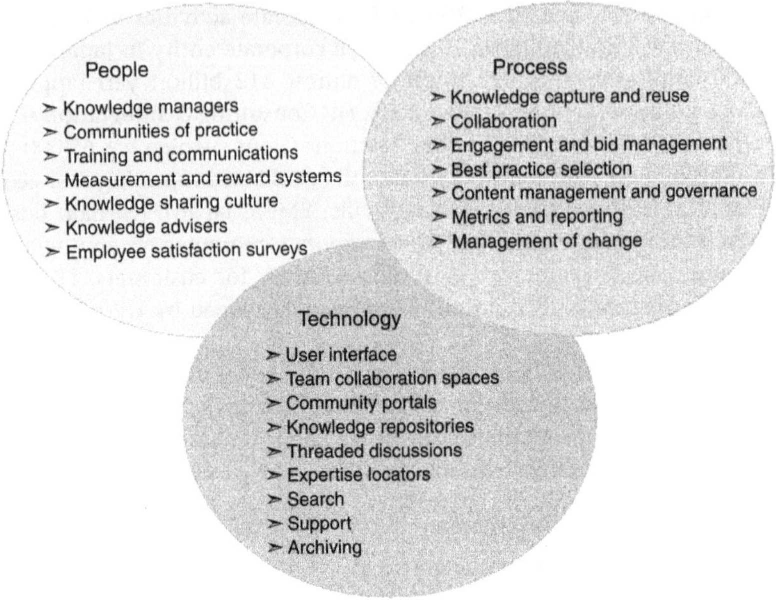

People
➤ Knowledge managers
➤ Communities of practice
➤ Training and communications
➤ Measurement and reward systems
➤ Knowledge sharing culture
➤ Knowledge advisers
➤ Employee satisfaction surveys

Process
➤ Knowledge capture and reuse
➤ Collaboration
➤ Engagement and bid management
➤ Best practice selection
➤ Content management and governance
➤ Metrics and reporting
➤ Management of change

Technology
➤ User interface
➤ Team collaboration spaces
➤ Community portals
➤ Knowledge repositories
➤ Threaded discussions
➤ Expertise locators
➤ Search
➤ Support
➤ Archiving

Figure 5.2 Knowledge management components at HP CI (from Kohlbacher and Mukai, 2007)

knowledge sources, based on their specific needs, and solicit feedback and utilize it for system improvements.

Moreover, community-based approaches to knowledge sharing and organizational learning are a key feature of HP CI's knowledge management activities. Offering not only hardware and software products but also a variety of IT and consulting services, HP CI's consultants and system engineers often work on different teams and different locations and thus need a location-independent and flexible solution for sharing their knowledge. In fact, as 75 per cent of the users are mobile, and many teams are geographically distributed, the web browser is the lowest common denominator for access for them. Generally at HP and consistent with CoP theory (cf. also 3.5), a CoP is a natural grouping of people who share and focus on a specific knowledge domain or topic, with the objective of creating, expanding, and exchanging knowledge, and developing individual and organizational capabilities. CoPs have no regional or organizational boundaries, live from their members' active participation and contributions, and offer a collaborative environment, discussion forums on topics of interest, as well as community building events (for example, HP Virtual Classroom). CoPs at HP are frequently referred to as learning communities, especially those that meet at regular teleconferences (cf. also Wenger, McDermott, and Snyder, 2002: 24) and recently this term has more and more been replaced by 'profession community'.

5.2.3 Knowledge-based marketing case: learning communities

5.2.3.1 Knowledge management at HP CI Japan

Japan is one of the four regions, along with the Americas, Asia Pacific, and EMEA, according to which HP CI is geographically divided. Japan occupies a special position within HP. In fact, Japan's particular ways of doing business and the specificities of the market and customers prompt a special approach in Japan. This is also true for the way people work and interact in organizations and the way they create, share, and disseminate knowledge. Indeed, research has shown that community building is also culture-dependent (Plaskoff, 2003).

As a result, HP CI Japan has applied HP's standard knowledge management activities only to some extent, and has adapted certain aspects, tools, and activities to their particular needs, ways of working, and sharing knowledge in Japan. Based on Nonaka's (1994) SECI model (cf. also 3.4 and 4.2.3), HP CI engages in knowledge management activities for capturing and leveraging its rich tacit knowledge base and encourages and supports the externalization and consequent re-use of this knowledge (the three main people-based activities). Additionally,

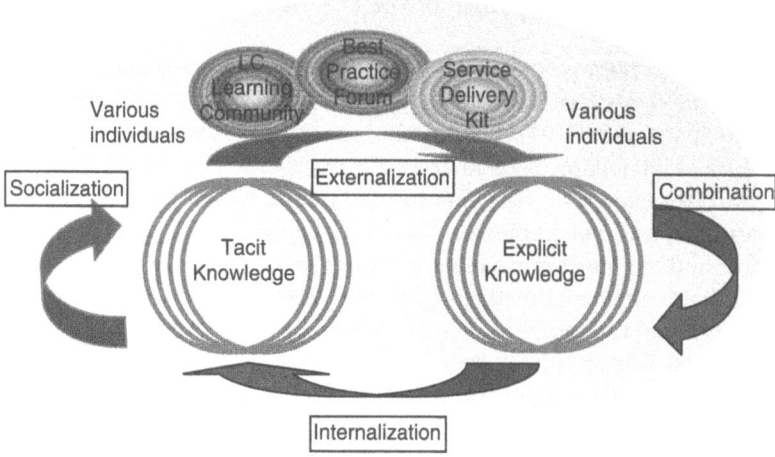

Figure 5.3   HP CI Japan's SECI model (from Kohlbacher and Mukai, 2007)

there are IT and tool-based knowledge management activities which not only foster the sharing of highly tacit knowledge but also help to make it explicit and thus easier for sharing and re-use. Figure 5.3 shows HP CI Japan's version of the SECI model.

The *Best Practice Forum* is an annual meeting for presenting, exchanging, and discussing success stories and best practices that have been achieved. It is held in the form of a competition and presentations, with material made available for all employees on the intranet. The *Service Delivery Kit (SDK)* is a collection of successful methods from experienced consultants with the aim of helping less experienced colleagues to learn and replicate approved practices to deliver superior service to HP's customers.

5.2.3.2 *Learning communities*

HP CI Japan's *Learning Communities (LC)* officially emerged in November 2001 from special interest groups (SIG) that had independently formed and worked in different departments. The SG business done by HP CI depends and thrives on the knowledge of individual employees. As all forms of consulting are people-based and people-centred, hence sharing tacit knowledge, externalizing and disseminating it, then the resulting explicit knowledge is essential for building and sustaining competitive advantage in the industry. In 2005, HP announced its education and career agenda, 'Profession Program', and as a result, LCs became part of the profession community, which requires mandatory participation for all employees. However, even within this new framework, the essence of the LCs has basically remained the same.

The main purpose of the LCs is twofold. First, the tacit knowledge of the individual consultants and system engineers is (partly) to be made explicit and shared, which is mostly done through discussions and professional interaction. Second, the LC is to provide a context and opportunity for executing HP's mentoring system, which is an important part of its internal employee education programme. All junior consultants and engineers have a senior counterpart assigned as their mentor who helps and supports them by giving advice and guidance. All in all, an LC's goal is to share knowledge and information about highly relevant and important issues, and to discuss and exchange opinions about them. Obviously, the consultants stay in close contact with HP CI's customers. To provide superior service and solutions, value needs to be co-created with them. Through this interaction with the customers, the consultants build up an immeasurably valuable tacit knowledge base about customer needs and the way to provide service and solutions that lead to higher customer value and satisfaction.

LCs are gatherings of all employees that have expert knowledge or who are simply interested in participating, learning, and discussing these topics. Therefore, topics and issues for discussion are various and may also change quickly. Besides, all employees are welcome to participate, regardless of their affiliation or position. LCs might centre around certain business areas, technological issues, or solution aspects. In fact, CoPs are not primarily about a product, function, or tasks, but rather centre on a specific knowledge domain (Soekijad, Huis in't Veld, and Enserink, 2004; cf. also Wenger, McDermott, and Snyder, 2002). Having experts and people with the same interests and the same need for solutions gathered to discuss matters in groups and face-to-face has proven very beneficial for leveraging and exchanging tacit knowledge and finally making it explicit, thus adding to the organization's common knowledge base and reducing its dependency on the individual. Especially in heated discussions, people will end up making their points very clearly and expressing their opinions, thoughts, worries and even complaints, quite straightforwardly.

LC meetings usually start with a presentation on interesting or urgent topics and issues and will be followed by discussions. The presentations as well as other materials are made available on the intranet not only to the LC members but to all CI employees. The same is true of summaries of the discussions and meeting minutes of the LC. Examples of LCs at HP CI Japan are communities focused on certain types of products, such as Linux, databases or security software, certain methods such as IT Service Management and Project Management, and also certain fields of business, such as financial services and networks.

HP CI Japan's LCs are guided and co-ordinated by the knowledge management department whose staff also serve as facilitators and advisers for the communities, as well as all other knowledge management relevant topics and questions. The knowledge management department is also responsible for the handling and organization of registration to the community, usually on an annual basis, and training and administrative work resulting from the execution and maintenance of the LC. LCs mostly meet once every two weeks or once a month and participation varies between five and forty people. Besides the face-to-face meetings, LCs also employ mailing lists and LC forums on the intranet for quick and easy access and exchange of information and explicit knowledge. The regular meetings and discussions of the LCs help employees to share current information, news on important issues, and their expert know-how on certain topics, as well as their experiences, success and failure stories, and best practices. Thus, LCs also provide a space and a context for the education of their members and for the solution of concrete problems, as well as their pro-active prevention.

Finally, the biggest difference between HP CI Japan's LCs and the worldwide communities lies in the number of participants and the focus on the type of knowledge. While the communities are meetings of a large number of employees and often take the form of seminars or training courses and as such focus rather on explicit knowledge and the combination of sharing and transferring it, LCs in Japan usually only have a small number of people and focus on the sharing and co-creation of tacit knowledge. In fact, even though the LCs are of course trying to externalize as much tacit knowledge as possible, they acknowledge that not all tacit knowledge can be made explicit and in that case concentrate on the exchange and sharing of this tacit knowledge without formalization and externalization.

5.2.4 Conclusion

This case study illustrates how HP CI, and specifically HP CI Japan, leverage a particular form of CoP to create and share both tacit and explicit customer and other marketing knowledge. As the case study – as well as the extant literature – reveals, HP in general has a very strong and sophisticated knowledge management in place, which skilfully combines the three main components of people, processes, and technology. The case study focuses on one peculiar knowledge management activity, which plays a crucial role in creating and sharing tacit knowledge and is therefore different from many of the other knowledge management activities. Indeed, all in all, an LC's goal is to share knowledge and information about highly relevant and important issues, and to discuss and exchange opinions about them. Needless to say, the most valuable knowledge and insights are often very tacit and context-dependent. Besides, as shown in Chapter 4.1.3.3, customer knowledge is a specifically valuable knowledge for firms and frequently tends to be rather tacit. Through interaction with them, the field engineers not only gain knowledge about customers but also from customers and about how to support them. And by delivering its services at the customer's premises, knowledge and value can also be co-created between HP and its customers. However, this newly created knowledge should also be fed back to the organization, that is, to HP, where it can serve as the basis for further co-creation within the firm. This is where the LCs prove to be a helpful and powerful tool. To sum up, this case study shows that face-to-face communication and interaction are necessary to share and leverage tacit marketing knowledge and points to the importance of a shared context such as a CoP to support this.

5.3 Schindler Elevator

This case study illustrates how Schindler Elevator took a knowledge-based approach to marketing in launching a new escalator product in Asian markets and effectively co-created and transferred both explicit and tacit knowledge between the escalator headquarters (competence centre) in Vienna and the Asia-Pacific headquarters in Hong Kong.[31]

5.3.1 Company information/background

The Swiss concern Schindler Lifts and Escalators Ltd. is the world's largest manufacturer of escalators and moving walks and the world's number two for elevators. The Schindler Group comprises two core businesses: Elevators & Escalators, which contributed 78 per cent of sales in 2005, and ALSO, an IT distributor in Europe. From humble beginnings in the 1870s, Schindler is today a global company employing nearly 40 000 people with an operating revenue of more than CHF 8.8 billion (elevators and escalators 6.9 billion) in 2005.

Schindler's headquarters are located in Switzerland but the Competence Centre Escalators – which encompasses R&D, production, product management, intra-group sales, sales support activities, and so on – is based in Vienna, Austria. The factory in Vienna is now the second largest – in terms of assembled units per year – of four factories in total, with Shanghai being the largest. Asia-Pacific headquarters – until 2005 – were located in Hong Kong, where Schindler entered in the 1920s, using Jardine Engineering as its representative. This relationship eventually evolved into the Jardine Schindler JV. Schindler's business involvements with Japan go back to 1935, when a local company with the name Towa Elevator was founded in Japan. In 1954, a stock corporation with the name Nihon Elevator emerged from it and in 1985 Nihon Elevator agreed to establish a strategic alliance with the Schindler corporation. Today it operates in Japan under the name of Schindler Elevator KK, employing some 350 people throughout its twelve branches. The Tokyo head office reports to the Schindler group's Asia-Pacific headquarters in Hong Kong. In October 2005, Schindler acquired the Japanese elevator company Mercury Ascensore, a leading independent elevator and escalator maintenance provider.

This acquisition enabled Schindler to substantially increase the network density of its maintenance portfolio.

5.3.2 Knowledge management initiatives/activities in general

Even though Schindler has knowledge management on its management agenda, there are few explicitly termed activities and no knowledge management departments or knowledge managers. As for the escalator business, the most prominent knowledge management tool is the Intranet Portal Escalators, whose core is maintained by the Competence Centre Escalators in Vienna. It includes news about the Schindler group, the escalator industry and competitors, and technical and sales related information and data. Schindler subsidiaries abroad maintain their own local site within the intranet, thus trying to make explicit and retain as much local information and knowledge as possible. In addition to the Intranet Portal Escalators, information and knowledge are shared through the Lotus Notes groupware, its e-mail client, and databases.

5.3.3 Knowledge-based marketing case: escalator product launch in Asia

5.3.3.1 Asian markets

The countries in the Asia-Pacific region have been achieving good growth in recent years, mainly led by China, the world's largest market for elevators and escalators. This rapid economic development resulted in increased levels of construction activity in most Asian countries. The exceptional growth in China was driven by large commercial construction projects and by residential construction, as well as by government-led infrastructure projects such as airports and subways.

According to company information, Schindler achieved another excellent reporting year in the Asia-Pacific region in 2005, reflecting the favourable market environment. Sales grew more rapidly than the market, thus strengthening Schindler's position in the region. Its success was particularly evident in China, Hong Kong, Malaysia, Vietnam, and Indonesia. Moreover, Schindler secured several large projects in Macau, thus creating a good basis for its future growth. In 2005, Schindler consolidated its leading position in the escalator industry through the launch of the new Schindler 9300 Advanced Edition (see below) and the opening of the world's largest escalator manufacturing plant in Shanghai, China, with an annual production capacity of over 6000 units.

Asian – and specifically Japanese – norms for elevators and escalators diverge widely from those in Europe owing to the very strict safety regulations due to the high earthquake risk and fire protection. In practice these regulations are subject to continual modification with the result that it is impossible to supply more or less standard products in this market sector to Japan. Consequently, almost all Schindler elevators and arrays of otherwise standard parts have to be developed and produced exclusively for the Japanese market, whilst escalators and their parts have to be slightly adapted as well – the standardized parts or elevators Schindler produces worldwide can hardly be sold in Japan. Furthermore, Japanese customers baulk at standardized products which are perceived not to accommodate the 'uniqueness' of Japanese conditions.

Thus Japanese regulations, on the one hand, and Japanese vanity on the other, made local manufacture very costly and inhibited innovation. Both factors put Schindler Elevator KK at a competitive disadvantage against local elevator and escalator manufacturers, who dominated the market. The challenge for the Japanese subsidiary of the Swiss company was to introduce a new escalator (the elevator segment offering little movement), which had been introduced into other Asian countries, but which would need to be introduced into Japan with appropriate adaptations.

5.3.3.2 *The new escalator product and its launch*

The new product was the newly developed Schindler 9300® Advanced Edition escalator. In 2004, the Competence Centre Escalators in co-operation with local subsidiaries had launched, first in Europe and the Middle East and subsequently also in other continents. Market introduction in Asia followed in 2005, and the product launch in Japan finally in the middle of 2006. The new product offers greater choice in terms of its design and features and can also be installed more rapidly. Besides, it offers superior quality and a reduction in (production and installation) costs which could partly be passed on to the customers. As a matter of fact, the worldwide elevator and escalator business is characterized by severe price competition, which, in most cases, leads to sales below cost, with losses being recovered and turned into a profit through maintenance and service. Therefore, cost was a particularly crucial issue in Asian markets, including the large leading market of China and Japan. One of the main applications of the Schindler 9300® Advanced Edition escalator is for use in shopping centres, although it is by no means limited to that.

For the introduction of the new escalator, it was decided to implement a new marketing strategy which would be both value- and knowledge-based. In contrast to the past, the customers' attention was drawn away from seeing the escalator merely as an independent piece of hardware. In fact, Schindler's value proposition is the offering of an integrated solution, that is, all services in relation to the product life cycle of the escalator and even beyond. Concretely speaking, this includes the planning of buildings and the positions and numbers of escalators, traffic flow analyses – particularly important for shopping centres – installation, maintenance, exchange, and disposal. This obviously also involves the co-creation of value and knowledge with building planners, architects, general contractors, shopping centre owners, and so on. As a result, a product as simple as an escalator can actually involve a range of different services and collaborations with customers, often encompassing highly sophisticated planning processes.

For the launch in Asia-Pacific, experts were dispatched from the Schindler Competence Centre in Vienna to Hong Kong on a two- to three-year assignment and their task was not to impose a marketing strategy on subsidiaries in Asia, but to evolve one on the basis of knowledge-sharing and co-creation. The senior expert and his team prepared a special training programme for all subsidiaries in the Asia-Pacific region and ran a product road-show in different countries. The subsidiaries hosted in-depth workshops on the new product features, discussed marketing strategy, assessed competition, and evolved USPs, which were distinctive for each country. The workshops lasted one or two days and were a forum of knowledge between the experts and cross-sections of local staff, especially those concerned with sales, marketing, and technical support. There were interactive PowerPoint demonstrations, role-plays, and discussion of case studies and the introduction of the above-mentioned Schindler Intranet Portal Escalators as well as various Lotus Notes databases.

The experts were at pains to deliver explicit and tacit knowledge on products and competition in a culture-specific way by conveying trust and confidence in the local managers. The process not only created new bi-directional lines of communication between the Asia-Pacific headquarters and its local subsidiaries, but it also motivated these managers, who in the past had never been much consulted, to discuss freely all aspects of their experiences as Schindler employees. In this way Schindler acquired fresh and revealing tacit knowledge about its business operations in Asia. Much of this hitherto undisclosed knowledge was so useful that it was fed back to the Competence Centre in Vienna, being of value not only to product development staff but also to personnel in research and development. A further benefit of these interactions was that it meant that a distinctive marketing strategy could be developed for each market. It now had very rich contextual knowledge which it could convert into specific business approaches as well as specific adaptations of product offerings.

5.3.4 Conclusion

This case study illustrates how Schindler Elevator took a knowledge-based approach to marketing in launching a new escalator product in Asian markets and effectively co-created and transferred both explicit and tacit knowledge between the escalator headquarters (Competence Centre) in Vienna and the Asia-Pacific headquarters in Hong Kong. Even though escalators as hardware are by themselves neither particularly sophisticated nor complex products, a lot of critical tacit knowledge was involved in the product launch. For this reason, direct interaction and face-to-face communications became necessary to transfer and recreate this knowledge, as well as to co-create new knowledge locally and feed it back to the headquarters in Vienna. Moreover, Schindler had rethought its value proposition and shifted its focus away from merely offering product hardware to delivering integrated solutions, that is, all services in relation to the product life cycle of the escalator and even beyond. This also involves the co-creation of value and knowledge with building planners, architects, general contractors, shopping centre owners, and so on. As a result, a product as simple as an escalator can actually involve a range of different services and collaborations with customers, often encompassing highly sophisticated planning processes. Moreover, the case also shows that different local contexts in different countries have great implications and that therefore such a global product launch cannot be simply conducted by headquarters in a unilateral way, but needs to be done in close co-operation with local staff in the subsidiaries. To sum up, the case reveals how knowledge transfer within MNCs should be seen as a process of jointly re-creating existing knowledge rather than merely transferring it unilaterally. Additionally, through direct interaction and face-to-face communication new knowledge can also be co-created and fed back to headquarters, thus benefiting the entire knowledge base of the firm.

5.4 Siemens

This case study presents a recently launched company-wide cross-selling and marketing knowledge sharing initiative in Siemens and illustrates how cross-functional, cross-divisional, and cross-regional collaboration is leveraged for value and knowledge co-creation resulting in superior value propositions to customers.

5.4.1 Company information/background

Siemens – headquartered in Berlin and Munich – is a global conglomerate and a powerhouse in electrical engineering and electronics. The company has around 461 000 employees working to develop and manufacture products, design and install complex systems and projects, and tailor a wide range of services to individual requirements. Siemens provides innovative technologies and comprehensive know-how to benefit customers in 190 countries. Founded more than 155 years ago, the company focuses on the areas of Information and Communications (COM), Automation and Control, Power, Transportation, Medical (MED), and Lighting. In fiscal year 2005, Siemens had sales from continuing operations of EUR 75.4 billion and income from continuing operations of EUR 3.058 billion. Figure 5.4 shows the Siemens portfolio and sales by business area.

Since its establishment in Japan in 1887, Siemens has a long history of providing its customers with a diverse range of innovative products and services through direct activities, joint ventures (JVs), and co-operation with Japanese companies. Today, the Siemens Group in Japan conducts business in a wide range of areas including information and communications, automation and control, power, transportation, medical, and lighting. In 2005, the Siemens Group posted sales of 110 billion yen on a consolidated basis, and had a total of 1900 employees. Siemens K.K., the official representative of Siemens AG in Japan, is responsible for regional co-ordination of the Siemens Group

Sales by business area

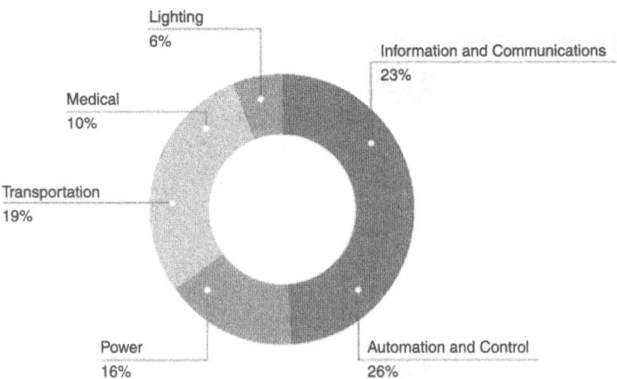

Figure 5.4 The Siemens portfolio (company information)

in Japan and in addition is responsible for business in the areas of communications, automation and drives, industrial solutions and services, logistics, power generation, power transmission and distribution, and transportation (mainly for railways). Placing special emphasis on the business of automation and drive systems, the Group established Yaskawa Siemens Automation & Drives Corp. and Yaskawa Siemens Numerical Controls Corp. Siemens Building Technologies K.K. designs and manages control systems for buildings. In the field of transportation, Siemens VDO Automotive K.K. was set up to handle car electronics businesses. Siemens-Asahi Medical Technologies Ltd. sells medical equipment such as diagnostic imaging systems, commanding a substantial share of the Japanese market. The hearing aids of Siemens Hearing Instruments K.K. are highly valued as well. In the lighting industry, OSRAM-MELCO Ltd. manufactures lamps meeting the needs of the marketplace and Mitsubishi Electric OSRAM Ltd. supplies these products to the market. OSRAM Ltd.'s main product line is automotive lamps. Moreover, in order to meet customers' financial needs in introducing Siemens products into Japan, Siemens Financial Services K.K. was established. Two other strategic alliances with Japanese firms are Mobisphere – a JV with NEC – headquartered in the UK and Fujitsu Siemens – Fujitsu itself actually originated from a JV between Siemens and the Japanese Furukawa group – headquartered in Germany.

5.4.2 Knowledge management initiatives/activities in general

Siemens is perceived as a pioneer and leading firm in knowledge management by both academics and practitioners (cf. also Davenport and Probst, 2002b). As such, Siemens has frequently been featured as a role model in numerous books and articles on knowledge management and knowledge sharing (for example, Davenport and Probst, 2002a; von Krogh, Ichijo, and Nonaka, 2000, to name just two bestsellers) and has also been a recipient of the MAKE (Most Admired Knowledge Enterprise) award several times (cf., for example, English and Baker, 2006). The 2004 European MAKE panel, for example, recognized Siemens for its 'enterprise knowledge-driven culture, developing knowledge-based products/services/solutions, and creating value from customer knowledge' (English and Baker, 2006: 207). Groups and regions of the company with joint co-ordinating and proficiency-building efforts practise knowledge management at Siemens using a decentralized approach (ibid.). Among the most researched knowledge management initiatives at Siemens are the COM division's (formerly ICN) 'ShareNet', and the MED division's 'KnowledgeSharing@MED'. Similar to the HP CI case above CoPs have proven a very effective and efficient forum for knowledge creation and sharing at MED. Additionally, through the interviews I have identified another interesting and effective knowledge management initiative at Siemens Building Technologies (SBT). It is called 'References@SBT' and is basically a best-practice case exchange forum and intranet-based platform for knowledge exchange.

5.4.3 Knowledge-based marketing case: Siemens One

5.4.3.1 Siemens One – bundling expertise for the customers

The Siemens One initiative is a company-wide strategy to improve market penetration and drive growth in new fields by enhancing co-operation across the entire organization. It is a key part of the Siemens Customer Focus programme, which is meant to optimize the partnership between the different Siemens groups and new and existing customers, while co-ordinating cross-group activities in key market segments. By strengthening co-operation across the vertically organized groups, Siemens aims at enabling customers worldwide to profit from its ability to combine a comprehensive array of innovative products and services in order to create complete, customized solutions. Offering their customers their products, services, and solutions from *one* source through Siemens One, and by bundling their capabilities, Siemens can offer more customers more innovation than before. In sum, 'the objective of Siemens One is to harness the power of Siemens for the optimum customer partnership and develop the whole company toward an even more customer-centric organization model' (Senn, 2006: 29).

There are now Siemens One organizations in over thirty-five countries and there is a Siemens One team at corporate headquarters to help the vertically organized businesses further leverage their horizontal synergies, to initiate cross-group and cross-region solutions in new and existing market segments, and to engage in sales activities where appropriate. The aim throughout is to optimize customer value by intensifying and expanding current cross-selling activities. Key account management (KAM) and sector management approaches further help to pool, arrange, and leverage relevant marketing knowledge. Similar to in-house consultants, Siemens One teams are the missionaries of Siemens' new marketing philosophy of customer focus and cross-selling activities and support the process of understanding, identifying, initiating, and implementing cross-group and cross-region customer solutions. An important part of this mission is the creation of a common vision of one Siemens and the establishment of a proper infrastructure for sharing information and jointly creating new marketing knowledge. This includes sales trainings, marketing intelligence activities, the collection and provision of corporate success stories, best practices and case studies, as well as the introduction of CRM tools and local Siemens One intranet portals that are connected to the global portal.

5.4.3.2 Siemens One Japan

In order to drive the Siemens One initiative in a systematic way, targeting to establish firm processes within the region, Siemens Japan started its Siemens One activities in June 2005 as an organization directly reporting to the CEO of Siemens K.K. First, a road map for the upcoming months was set up together with the Corporate Development Siemens One team at corporate headquarters in Munich. Increasing cross-selling activities and winning new customers within and across business groups is the goal of the operational excellence initiative at Siemens and the idea of Siemens One is to enable all business groups to act as team players based on a defined account management and sector development methodology.

Supporting the business groups to act as one Siemens by applying a systematic customer perspective and challenging them to use the methods and concepts developed by the customer focus initiative is the mission of Siemens One. In order to maximize the respective group profit, market transparency is a prerequisite to set up the right strategy and increase market share through leveraging market-fit products and solutions in accordance with customer requirements. Siemens can offer a wide range of products and solutions (power plants, communications equipment, automation and drive controls, building technologies, and so on) to the customers in their sectors to satisfy their individual needs. As a result, customers can reduce the complexity in large-scale projects and multi-vendor environments with the help of Siemens.

One important task for Siemens One is the implementation of a sector management approach. Siemens One defines key relevant industry segments such as automotive, health care, airports, oil and gas, semiconductors, cement, and so on, as industrial sectors in order to drive the products and solution business in a systematic way. The so-called Sector Development Board (SDB) – a management body to strengthen the customers' high acceptance of Siemens, and achieve highest customer satisfaction – has set up sector strategies on a global basis and regional companies are a part of this global sector strategy. Taking this sectoral approach means Siemens offers comprehensive tailored solutions through thorough investigation of customer needs as well as sector-specific requirements to the solutions. SDB is also a decision board for the sector's global strategy, overall management for marketing concepts, resources and solution portfolio. The Sector Support Team (SST) is responsible for driving the regional business through close co-operation with regional Siemens One teams and regional business units. Siemens One Japan analysed the activities of the Siemens Japan group and business units according to the sector definition, and set up four first sectors in Japan (as of July 2006). The established sector team members commonly approach selected key accounts by identifying business opportunities. Obviously, sharing of existent as well as the creation of new markets and other marketing knowledge, specifically from different perspectives – different functions, different business areas – are crucial here.

Another essential task for Siemens One is to establish an account management system. Siemens account management is a vital part of the corporate CRM process. The CRM objective is to maximize the level of customer satisfaction and develop the corporate, international, and national top accounts in a systematic way. Through optimizing the relationship between markets/customers and Siemens, a long-term relationship with strategic customers is built (cf. also Senn, 2006). The sales organization is the key function for implementing CRM, and they have the CRM roll out responsibility for their accounts. Moreover, a visualization system for the groups and regions is a key issue for successful account management, which is a success factor for the Siemens One activity at the same time. In order to gain transparency for account management, account information (customer project information, account penetration plan, technology requirements, and so on) must be shared by the team members who are involved in account management. For that purpose, Siemens fully utilizes the ATP (Account Team Portal) as a cross-group and regional CRM tool. In this portal, team leaders and team members regularly update customer information, business opportunities, and business development data which are thought to be critical in order to drive account management towards success. All the stored data are then utilized by the team members in Account Planning Process Plus (APP+), a common account management workshop. Sector managers, team leaders, and their members then conduct a systematic account management meeting to identify business opportunities, Siemens offerings, and to define a common sales strategy. Account teams repeat these APP+ processes as regular business opportunities workshops. One of the benefits of this APP+

workshop is that all account team members are able to have the same recognition of customer requirements, business potentials, Siemens solutions, and common strategy with clear individual roles and responsibilities.

For both sector management and account management, market intelligence and market research activities are essential. Therefore, the creation of market transparency is also high on the agenda of Siemens One. At Siemens it is assumed that once 60 per cent market transparency and understanding can be achieved, the market share can be increased by more than 10 per cent ('60/10' rule), for example. Generating and collecting the needed information and knowledge is also necessary to implement standardized processes for this and for the presentation and reporting of the results. Moreover, transparency is also needed within Siemens, which requires regular information and knowledge sharing and reporting to each other.

Last but not least, the systematic creation, collection, arrangement, and sharing of best practices throughout functions and divisions is also an important task of the Siemens One initiative. Each business unit and group company in Japan has its success stories, and the Siemens One Japan team conducts common success story/best practice workshops to share tacit and explicit knowledge of experienced employees. Besides, there is also a newly created customized intranet portal in both Japanese and English to present Siemens One, account management, sector management and other relevant concepts and processes. The intranet portal also hosts local sector portals for information sharing and exchange with sector teams in other countries. Figure 5.5 summarizes the key Siemens One activities in one diagram.

5.4.4 Conclusion

This case study presents a recently launched company-wide cross-selling and marketing knowledge sharing initiative in Siemens and illustrates how cross-functional, cross-divisional, and cross-regional collaboration is leveraged for value and knowledge co-creation resulting in superior value propositions to customers. It is a very comprehensive case of knowledge-based marketing in action as it not only spans the boundaries of different business divisions within the MNC but also transcends the boundaries of the firm and involves customers, partners, and so on (cf. also 6.2). As Siemens One was born from and is driven by the customer focus initiative within Siemens it is important to mention and discuss it, but the main focus in this book is on the

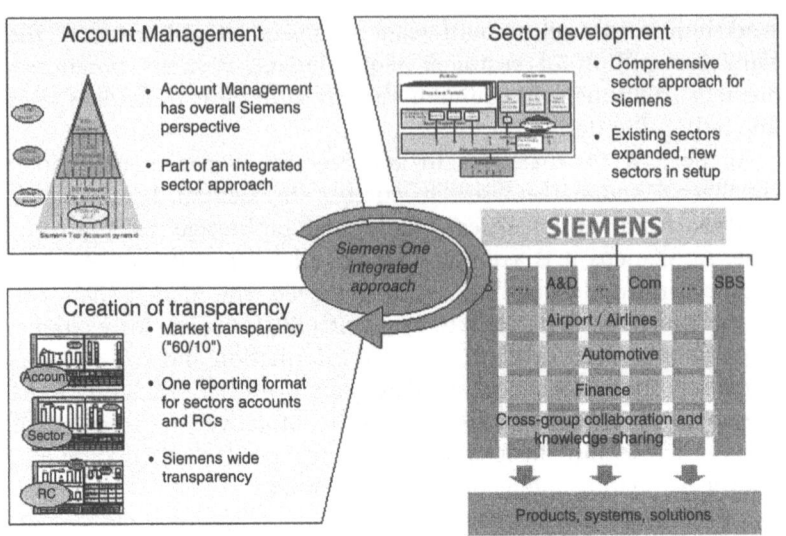

Figure 5.5 Key Siemens One activities (based on company information)

co-creation of knowledge and value within the MNC. Here the role of both tacit knowledge and of the 'in-house consultants' or facilitators – 'knowledge activists' (cf., for example, von Krogh, Ichijo, and Nonaka, 2000) – are highlighted. Sharing of existing knowledge as well as the co-creation of ideas and new knowledge through cross-functional, cross-divisional, and cross-regional collaboration is critical for providing superior products and services and for the way existing customers are served and new ones acquired. Moreover, it became obvious that such an approach needs to be planned, designed, and initiated strategically at headquarters, but rolled-out, implemented, adapted, and maintained locally in the subsidiaries, but still with support from headquarters. To sum up, the case illustrates the strategic importance of marketing knowledge and its management – creation and sharing – throughout the company. It also underlines that marketing knowledge does not reside merely within one corporate function but that it is dispersed throughout different functions and divisions and can only be properly leveraged through active collaboration.

5.5 Toyota Motor Corporation/Toyota Peugeot Citroën Automobile Czech

This case study illustrates how Toyota leverages marketing knowledge through a JV with a competitor in order to expand its business in

Europe and crack the East European market and also co-creates new knowledge with the alliance partner.

5.5.1 Company information/background

Toyota is Japan's largest manufacturer of automotive vehicles and probably the world's most successful one. On consolidated accounts, Toyota employs more than 285 000 people and has net sales of 21,036.9 billion yen (fiscal year 2005). The Toyota Motor Corporation has built a strong reputation for the high quality, durability, and reliability of its cars, and these are only some of the reasons for its outstanding global success. Research and academic writing on Toyota is voluminous (for example, Dyer and Nobeoka, 2000; Liker, 2004; Sobek II, Ward, and Liker, 1999; Spear and Bowen, 1999; Womack, Jones, and Roos, 1991, to name just a few) and Toyota frequently serves as a role model for both academics and business practitioners. Indeed, the teachings of the so-called 'Toyota Way' and the legendary Toyota Production System (TPS) (for example, Dyer, 1994; Fujimoto, 1999; Kamath and Liker, 1994; Liker, 2004; Pine II, Victor, and Boynton, 1993; Sobek II, Liker, and Ward, 1998; Sobek II, Ward, and Liker, 1999; Spear and Bowen, 1999; Ward, Liker, Cristiano, and Sobek II, 1995) have not only been applied to manufacturing and production but also to other areas such as health care, postal services, and the service industry in general (Liker, 2004; Spear, 2004, 2005; Womack and Jones, 1996, 2005a). In fact, the popularized version of TPS, lean management, and lean manufacturing – starting with *The Machine That Changed the World* (Womack, Jones, and Roos, 1991) – have become an extremely successful 'management fad' lasting up to the present (Womack and Jones, 1994, 2003, 2005b).

5.5.2 Knowledge management initiatives/activities in general

Toyota has often been found to be very strong at organizational learning and knowledge creation and sharing (cf., for example, Ichijo, 2006a; Ichijo and Kohlbacher, 2007; Liker, 2004; Spear, 2004; Spear and Bowen, 1999). Indeed, for Liker (2004: 13, xv), 'Toyota is a true learning organization that has been evolving and learning for most of a century' and thus has created 'one of the few examples of a genuine learning enterprise in human history'. Dyer and Nobeoka (2000: 346) seem to agree when they contend that 'Toyota, in particular, is widely recognized as a leader in continuous learning and improvement'. One aspect that has particularly been under the scrutiny of researchers is knowledge sharing and learning within its supplier network and the way Toyota leverages this co-created knowledge for both itself and its suppliers (for example, Dyer and Hatch, 2004, 2006; Dyer and Nobeoka, 2000; Dyer and Ouchi, 1993; Evans and Wolf, 2005; Liker and Choi, 2004). But rather than the mechanical 'management' of knowledge, it seems to be Toyota's peculiar corporate culture that enables continuous organizational learning and thus has made Toyota become a true learning organization.

5.5.3 Knowledge-based marketing case: Toyota Peugeot Citroën Automobile Czech (TPCA)[32]

5.5.3.1 TPCA: Joint forces in the Czech Republic

Toyota Peugeot Citroën Automobile Czech (TPCA) is an IJV between Toyota Motor Corporation and PSA Peugeot Citroën in Kolín, Czech Republic. Both companies own exactly half of the shares (50/50 JV). After an agreement in July 2001, on 9 January 2002 the two automakers announced, the signing of an official JV agreement to establish TPCA. It is Toyota's sixth manufacturing company in Europe. The TPCA factory alone employs about 3000 Czech employees, and indirectly ensures an additional 7000 jobs in all areas, from the production of automobile components to cleaning services.

5.5.3.2 The East European challenge/the East European shift

With this unique automobile partnership and its joint plan for the development and production of small compact vehicles and the construction of a new factory, Toyota and PSA decided to react to the changing European customer market and to found a whole new category of small modern and technologically advanced vehicles. In fact, both companies see growing demand for such cars in Europe, and the new-platform vehicles to be built in the Czech Republic will be marketed under the Toyota, Peugeot, and Citroën brands. Total investment in the project on the grass field – including R&D and business startup costs – has surpassed 50 billion crowns (approximately 1.5 billion euros) and finally started manufacturing on 28 February 2005. The plant manufactures 300 000 small gasoline and diesel cars annually to be sold in Europe under both automakers' brands, that is, 200 000 units for Peugeot and Citroën and 100 000 for Toyota. The three, all-new small cars produced on a common platform are: the Citroën C1, the Peugeot 107, and the Toyota Aygo.

5.5.3.3 Three models – one platform

The cars built on this new platform have been developed jointly by the two companies. They are a modern, four-seat model boasting the most sophisticated technologies in terms of safety, reliability, environmental protection, and urban mobility. Equipped with the latest generation of 1.0-litre gasoline engines and 1.4-litre diesel engines, they are especially fuel-efficient. The project offers clearly differentiated models and specific styles for the vehicles of the two groups while guaranteeing strong commonality for the car structure and components. In launching this new vehicle concept, Toyota and PSA have introduced a brand new offer of small-size cars which will complement their product lines. This decision to jointly introduce a new class of cars, positioned below current entry-level models, is in response to changing needs in Europe, a market where demand for compact vehicles remains as strong as ever and is forecast to increase in the years ahead. Therefore, TPCA paves the way for a new market of vehicles, which retain all the essential features of a real car, and offer, at attractive prices, efficient solutions to environmental and urban mobility-related requirements. Target customers also include current buyers of used or outdated cars. In fact, primarily designed for European markets, this new car concept has been conceived to meet the changing needs of local customers. Cars produced using this common platform have a threefold advantage: they have prices lower than those in the current small-car segment, feature a high-level of standard safety performance, and offer excellent environmental features.

5.5.3.4 The best of both worlds: get success

The joint production at TPCA not only allows for a reduced cost but also a connection of the best of both automobile factories: the untouched production system of Toyota with the excellent knowledge of the European market of PSA. Therefore, Toyota is in charge of development and production while PSA is responsible for procurement. Toyota's Polish plant – Toyota Motor Manufacturing Poland Sp. (TMMP) in Walbrzych, Poland was established on 7 June 2002, as Toyota's first European transmission plant – will expand to supply manual transmissions and 1.0l gasoline engines for the Czech plant. PSA Peugeot Citroën will supply 1.4l diesel engines. Nearly all other components will be sourced locally. In fact, since the establishment of the JV, many Toyota-affiliated parts makers have set up shop in Central and Eastern Europe, and about twenty have signed supplier agreements with TPCA. PSA views this co-operation between two independent companies as a further materialization of the PSA Group's strategy aimed at reaching agreements on the joint development and production of mechanical components and platform elements, with the objective of obtaining economies of scale.

5.5.3.5 *Inter-organizational knowledge creation and sharing*

The TPCA plant is the fruit of a successful co-operation project that allowed the two global carmakers to combine their knowledge of product design, styling, production, and supplier relationships, while learning from each other's corporate cultures, technologies, and processes, as there seems to be very good teamwork and co-operation between Toyota and PSA. This led to an exchange of a wealth of specific know-how: PSA's knowledge of small cars in Europe and its expertise in purchasing activities, and Toyota's skills in development, manufacturing, and production processes. Indeed, Toyota learns from PSA mostly about purchasing issues, supplier relationships, both from a European and a general point of view, and even about production methods and shift management. On the other hand, PSA learns about Toyota's management style and the Toyota Way, and TPS and the production process. This mutual learning also leads to new joint ideas and knowledge co-creation.

As of March 2006, there were thirty-eight expatriates employed at TPCA: twenty-nine from Toyota and nine from PSA. All expatriates have management functions and include the President (Toyota) and Vice-President (PSA). They usually stay in the Czech Republic for about one and a half years before returning to their respective headquarters. Local staff are sent to Toyota's other European plants (Turkey, UK, France), and in the case of managers also to Toyota headquarters in Japan, for training for a period of one to six months. This ensures that they directly and interactively learn the Toyota Way and that local knowledge from TPCA can also be transferred back. Besides, over time, responsibility is handed more and more over to local staff after they have been trained by experts from Toyota and PSA.

5.5.3.6 *Excursus: Innovative International Multi-purpose Vehicle (IMV) Project*

Initially, Toyota developed and produced cars only in Japan and exported them abroad in order to ensure high quality and to maintain customer trust in the brand. Then, because of increasing overseas demand, the need to tailor production to local needs, the opportunity of tax breaks, and in order to save shipping costs, Toyota started to produce vehicles where the market is. This model has been working well in established mass markets such as North America and Europe, but recently, Toyota has identified attractive business opportunities in other developing markets, such as BRICs (Brazil, Russia, India, and China). The solution for globally operating companies – including

Toyota – has in the past tended to be to build manufacturing facilities in developing markets mainly owing to their cheap labour costs. Nevertheless, in developing and producing cars for these regions, Toyota used to stay reliant on Japanese designers and engineers, rather than exploiting local talent.

In 2004, Toyota announced a breakthrough initiative called the 'Innovative International Multi-purpose Vehicle (IMV) Project', which aims at increasing the self-reliance of overseas manufacturing facilities in such a way as to optimize overall worldwide production, especially in emerging markets, by both understanding common needs and paying sufficient attention to unique local needs (cf., for example, Ghemawat, 2005; Ichijo and Kohlbacher, 2007). The initiative is led by Toyota's subsidiaries, and in this business model, Toyota upgraded and expanded plants in Thailand, Indonesia, South Africa, and Argentina. This project is now dependent upon close collaboration between Toyota in Japan and its subsidiaries in emerging markets.

Within these emerging markets, the study of unique local needs and then the developing, manufacturing, and supplying of cars which closely meet them promises competitive advantage. 'Learn local' is the key to local success. But there is a global dimension, too. IMV cars assembled in Thailand and Indonesia are both used for local consumption and exported to different countries, particularly emerging markets. Surplus IMV cars assembled in Argentina are exported to Central and South America, and those assembled in South Africa are shipped to the rest of Africa. This global, cross-country collaboration is another key to the success of the IMV project. While paying attention to local unique needs in each region, Toyota tries to accomplish effective use of resources worldwide to provide high quality cars at lower cost. 'Act global, learn local' is thus another winning formula for the IMV project.

5.5.3.7 *Learn local, act global*

Supported by their spirit of teamwork and reinforced by a favourable national environment in terms of solid industrial experience and a quality education system, TPCA completed all stages of the co-operation successfully in terms of deadlines and results. This co-operation between independent carmakers has provided a fast, cost-efficient response to market demand through the sharing of expertise and experience. As a matter of fact, leveraging synergies and fostering mutual knowledge sharing and creation between the two partners are two of the most important goals and merits of this strategic alliance. Here, Toyota consistently follows its new 'learn local, act global' strategy by feeding back the newly created and acquired knowledge to its headquarters and spreading it also to other subunits.

Figure 3.7 in Chapter 3 shows different levels at which knowledge transfer and co-creation can occur in IJVs and Figure 5.6 illustrates it for the concrete example of TPCA. Obviously, knowledge will not only have to be created locally at TPCA but also between the JV (child) and the partner firms (parents). Moreover, the process of transferring newly created knowledge at TPCA to the parents is also critical. Inkpen (1998: 79) goes so far as to state that '[w]ithout active parent firm involvement in the alliance learning process, learning will not occur'.

In early 2006, the Toyota Aygo was doing extremely well on the forecasted resale values of the major leasing companies and the launch was widely seen as successful (Rädler, 2006). Unlike the mature market of Western Europe, the Eastern European automobile market offers much room to grow, with an automobile diffusion rate about half that of countries belonging to the European Union (EU). On joining the EU in May 2005, the five Central European nations have become subject to the bloc's tighter environmental regulations, so new car demand will likely increase as older cars are scrapped. As a result, Central and Eastern Europe will be vital for Toyota if it wants to achieve its goal of selling 1.2 million units a year in Europe by 2010. TPCA will help to

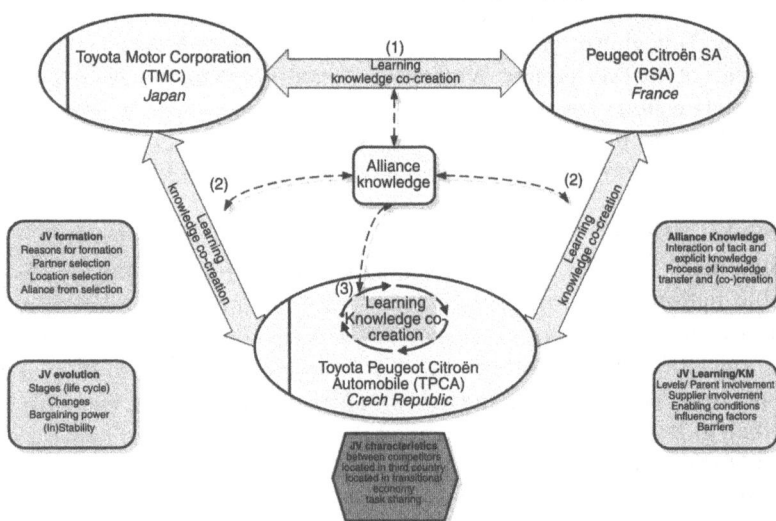

Figure 5.6 Levels of knowledge co-creation at TPCA (author's own illustration)

build a strong production base and will serve Toyota as a springboard for expanding its presence in Europe.

5.5.4 Conclusion

This case study illustrates how Toyota leverages marketing knowledge through a JV with a competitor in order to expand its business in Europe and crack the East European market and also co-creates new knowledge with the alliance partner. Similar to HP and Siemens, Toyota is frequently featured as a role model in knowledge management and organizational learning. But differently from the former two companies, in Toyota's case it is its unique organizational culture and the way the whole organization ticks rather than specific sophisticated methods or tools. Interestingly, Toyota seems to have been successful in 'exporting' and recreating this unique organizational learning culture also in foreign markets and with local staff. In a more difficult situation, the Toyota way of knowledge creation has to be applied to a new setting where two different corporate cultures as well as different national cultures meet. Toyota once successfully mastered such a situation in NUMMI with GM and now it is TPCA with PSA. But in the former case, Toyota was basically in charge of everything, while TPCA is a real collaboration between two equal partners. In order to mutually tap into each other's unique tacit knowledge base, direct interaction and face-to-face communication are necessary. At TPCA, this rich tacit knowledge comes from three sources: Toyota expatriates, PSA expatriates, and the local staff. To sum up, the TPCA case study together with the background information on the IMV project illustrate the new Toyota approach to global knowledge creation. Indeed, it strategically selects and leverages internal as well as external knowledge sources and engages in the co-creation of knowledge and value with its subsidiaries as well as other entities in the business ecosystem, such as competitors and suppliers.

5.6 Mazda Motor Corporation

This case study illustrates how an essential product concept and related tacit marketing knowledge were passed on and refined over three product generations. It also shows that the experience of co-creation between the product and the customers is crucial and that empathic design to tacitly understand customer needs is vital.

5.6.1 Company information/background

Originally established in 1920, Mazda started manufacturing tools in 1929 and then branched out into the production of trucks for commercial use before it launched its first passenger car models and began developing rotary engines. As of fiscal year 2005, Mazda Motor Corporation employed more than 36 000 people and its net sales amounted to almost 3000 billion Japanese yen. Mazda's largest stakeholder is Ford Motor Company.

5.6.2 Knowledge management initiatives/activities in general

Mazda's strength in product development, the creation of product concepts, and knowledge within the process have been documented by Nonaka and Takeuchi (1995) for the development of the rotary engine and the RX-7, and Mazda's product development in general by Cusumano and Nobeoka (1998). Mazda recently established a knowledge manager but knowledge management initiatives still seem to be at an initial stage. At the moment, the focus is on web-based communities and Q&A-sites on the intranet – the whole system is called 'Knowledge Wonderland' – for bringing the customer's voice into the company through salesmen and dealers and sharing important information and knowledge.

5.6.3 Knowledge-based marketing case: product development of the Mazda MX-5[33]

5.6.3.1 The Mazda MX-5 (Miata)

Mazda Motor Corporation unveiled the Mazda MX-5 (Miata) for the first time at the Chicago Auto Show on 9 February 1989. Sixteen years later, in 2005, the global debut of the all-new, third-generation Mazda MX-5 at the 2005 Salon International de l'Auto in Geneva demonstrated that producing a simple, lightweight sports car is fundamental to the soul of Mazda. In May 2000, the *Guinness Book of World Records* had recognized the Mazda MX-5 as the best-selling two-seater convertible sports car in history, with 531 890 units produced to that date. Since that affirmation, demand has held strong and more than 700 000 Mazda MX-5s have been sold around the globe, thus breaking its own record in 2005. In order to renew the MX-5, Mazda engineers focused on evolving the lightweight sports car concept as the all-new car had to follow in the tracks of a modern motoring icon. Finally, the Hiroshima-based company has successfully managed the challenge of developing and evolving the classic concept of the two-seat roadster and the new Mazda MX-5 was voted car of the year in 2005/2006 in Japan.

5.6.3.2   Rider and horse as one – Jinba Ittai

In Japanese, the exceptional soul of the Mazda MX-5 is described by the expression *Jinba Ittai*. The direct translation of the idiom is 'rider and horse as one' and can also be interpreted as 'oneness of man and machine'. *Yabusame*, a long-standing artistic ritual ceremony in Japan, truly embodies the essence of *Jinba Ittai*. An archer mounted on horseback gallops past a target and shoots an arrow. To achieve a bull's eye, the archer and horse must move as one. There must be a natural two-way communication and a high degree of synergy in their alliance. This oneness of motion between rider and horse was selected as the most apt analogy depicting the relationship between the driver and a Mazda MX-5. Updated for the twenty-first century, *Jinba Ittai* is akin to the bond between a single-seat Formula One driver and his car. It is also exemplified by a high-performance sports motorcycle rider at speed.

The rider-and-horse idiom and the effort to create a car universally seen and experienced as 'lots of fun' – the sub-concept of the roadster – served as the focal point around which the original and the all-new Mazda MX-5 were designed and engineered. While most sports cars aim for specific performance targets – such as the time required to accelerate to 100 km/h or cornering G provided by the chassis – Mazda engineers established additional goals to reinvigorate the lightweight sports car. In essence, this became a celebration of the simple delights of driving an open roadster. The 'fun' was designed for anyone and any location during sports driving and daily life.

5.6.3.3   Heritage and evolution of Jinba Ittai

The concept of *Jinba Ittai* was thought of and developed by the project leader of the first generation roadster, Toshihiko Hirai. He was inspired by his experiences in the field as a service engineer, when some of the dealerships and even customers questioned Mazda's *raison d'être* by asking: 'Why is there a need for cars from Mazda when we already have the same cars from Toyota and Nissan?' Hirai thought such inquiries endangered Mazda's right to exist in the market. Indeed, at that time – the middle of the 1980s – nobody believed there was a market or a necessity for a lightweight sports car. It was then that Hirai started to seek a fundamental solution in everyday business operations as a way of life. Confronted with this stark reality on the ground, he intuitively sensed for the first time a potential need for a lightweight sports car. This hunch soon turned into a conviction which led him to take the risk and take up the challenge to go for the innovative development. Finally, Hirai – in a bold effort and with hard work – managed to overcome management and other people's doubts about the need and demand for the roadster. With basically no budget or time officially assigned to the project at the beginning, Hirai still succeeded in building an excellent product development team.[34] The problem, however, was how to realize and implement the concept of *Jinba Ittai* in development and finally also the car itself. In order to convey his ideas and the meaning of *Jinba Ittai* for the car, Hirai experimented with *Kansei* engineering (see below) and used a fishbone chart (Figure 5.7) – as commonly used in TQM – to depict six key categories that guided the effort towards the desired 'oneness'. They are styling (inside and outside), touch (every aspect concerned with the tactile sense), listening (dominated by the engine's voice but also encompassing wind effects), cornering (handling dynamics), driving (everything from ride quality to acceleration response), and braking. Each of these categories was further broken down using the fishbone chart technique.

Takao Kijima was Hirai's successor for the development of the second generation roadster (launched in 1998) which carried on the concept and basic design of the first generation and thus incorporated only minor changes in the car, but was also very successful. As Kijima was already over fifty years old, he – at first – did not have any intention of serving as project leader for the third generation roadster. But when the plan for this car was revealed in 2001 he suddenly had a sense of crisis, especially as he realized that this project was more of a challenge than the last one. First of all, for the sake of efficiency, it was planned to share the platform with another, bigger car, which meant an increase in both weight and price (this urge to cut costs and share platforms came mainly from Mazda's major shareholder Ford). Kijima felt this

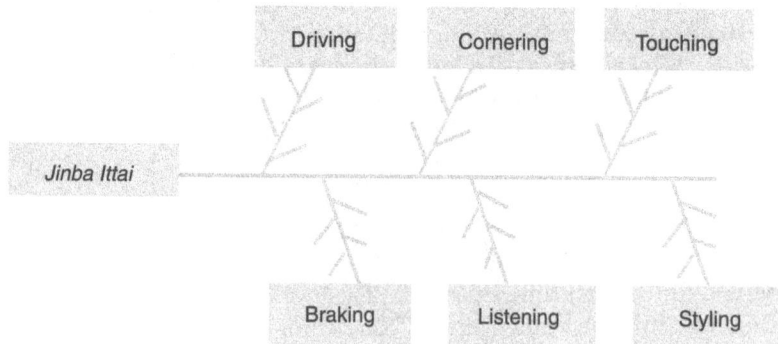

Figure 5.7   *Jinba Ittai* fishbone chart (based on company information)

would destroy the original identity of the roadster. Second, due to reinforced safety and environmental regulations, all parts and components would have to be changed. In the end, Kijima decided to take up the responsibility and lead the product development a second time.

Kijima had already served on the first generation roadster product development team under Hirai and had been in charge of chassis development. Thus, he knew Hirai's ideas, the product concepts, and the product itself very well, especially since he had led the development of the second generation, too. But for the third generation development, the circumstances and preconditions were completely different. Due to retirements and other assignments, Kijima had to work with a totally new team. This meant that he could not count on the tacitly shared knowledge of the team members. Indeed, except for the fishbone charts, hardly any of the tacit knowledge concerning ideas, the product concept, and so on, had been externalized and documented and thus had remained tacit and vanished with the leaving team members.

The first step for the engineering team was to acknowledge that *Jinba Ittai* was what made the original Mazda MX-5 so attractive for sports car enthusiasts and one of Mazda's greatest success stories. The second step was to use modern engineering methods developed by Mazda to evolve the iconic Mazda MX-5 into an all-new, third-generation edition. To convert the subtle imagery of *Jinba Ittai* into nuts-and-bolts reality with a body shell, a drivetrain, and chassis components, the engineers wielded a tool called *Kansei* engineering (cf. also Kijima and Hirai, 2003), which had already proved successful for the development of the first roadster under Hirai.

Like *Jinba Ittai*, *Kansei* is difficult to translate into Western terms. 'Thoughtful awareness' and 'heightened sensitivity' are the expressions that come closest to defining *Kansei*. It is the realization that the fitness of every constituent part underlies the goodness of the whole, that the aforementioned synergy between driver and car can be honed so that it can be felt by anyone who experiences the vehicle in motion. Another *Kansei* canon is that every aspect of design, mechanical function, and dynamic response ultimately contribute to driving satisfaction.

So, with *Jinba Ittai* as the point of origin and *Kansei* engineering as the navigation tool, as well as another set of new fishbone charts, the project team created an all-new Mazda MX-5 that is markedly better than, but not conceptually different from, the original. In order to better understand the special characteristics of the roadster and to directly and physically experience the product and its concept, all forty-two component team leaders extensively test drove all kinds of lightweight sports cars, including the first two generations of the Mazda roadster and the most renowned sports cars in the global market (for example, from Honda, Porsche, Fiat, BMW, and so on). All of them would go on long drives during the day, and then share their experiences, feelings, sensations, and thoughts over drinks together at night, often in heated and emotional discussions. They called this test driving and opinion and sensation sharing their 'concept trips'. These personal interactions and dialogues not only fostered a strong team spirit but also harnessed the recreation and sharing of a joint understanding of the concepts of *Jinba Ittai* and 'lots of fun'. It helped the team members generate a joint understanding of the essence of an ideal sports car, its aesthetics, quality, and motion.

Another important step was when all the team members were asked to describe and explain their interpretation of *Jinba Ittai* by writing it down in a small booklet – called the 'concept catalogue' – which subsequently served as the basis for the engineers' discussions. Creating this concept catalogue not only proved helpful for brainstorming and the arranging of ideas, but also created a kind of sense of commitment among the team members – a sense of 'self in the whole'. The concept catalogue summarized each person's general view of *Jinba Ittai* as well as how to achieve and realize *Jinba Ittai*, in the respective area of

Idea/view of *Jinba Ittai*	How to achieve/realize
The most important characteristic of the roadster is the concept of lightness, and this feeling has to be conveyed not only in driving but in every function and in performance overall. The feeling of *Jinba Ittai* has to hit the driver as soon as she gets into the car and within the first 10 miles of driving. If the notion of driver and car as one is only understood after driving the car for a period of time, or because the concept has been explained in advance, the roadster's existence cannot be justified. *Jinba Ittai* has to be experienced immediately. This is what makes the roadster what it is.	Rather than simply using lap time and torque chracteristics benchmarks for development, as was done in the past, development of the roadster has to focus on the whole car and then target its composite characteristics. The relationship of speed to G-force, sound of the engine, and control, is already understood. What we need now is the numerical breakdown to actualize the theme of 'Linear & Lively' in the P/Feel5 axis to create an outstanding ride.

Figure 5.8 Extract from the concept catalogue (based on company information)

responsibility of the team member. Figure 5.8 shows an example of one engineer's idea of *Jinba Ittai* and the means of implementing it in the development of certain parts of the car. In addition, all critical processes were documented by also using videos, pictures, and so on. The team members believed they were 'creating history' and were therefore eager to conserve as much as possible for future projects and different teams.

5.6.4 Conclusion

This case study illustrates how an essential product concept and related tacit marketing knowledge were passed on and refined over three product generations. It also shows that the experience of co-creation between the product and the customers is crucial and that empathic design (cf. also 4.1.3.2 and 4.1.3.4) to tacitly understand customer needs is vital. Note the location and point of time of value creation here. First of all, the value is not merely created within the factory or the firm, but in using the product, that is, driving the car. But it is even more than that because simply driving the car would not be enough. It is about the experience the customer has when driving the car, the feeling of *Jinba Ittai*. When the customer feels *Jinba Ittai* for the first time, the experience and thus value is created for the first time. The experience and the value are actually co-created between the customer and the product and it was empathic design and knowledge creation and re-creation in Mazda that made this possible in the first place. Here, the power of language, of images and metaphors to convey customer needs into product concepts was very important. Indeed, it is this kind of high-quality tacit knowledge – holistic knowledge – that becomes the source of innovation, and skilfully and empathically applying this knowledge for creativity will finally harness real innovation. The importance of the product development project leader and his decisive role also need to be pointed out. In fact, he served as a knowledge activist as well as a motivator and driver for the whole team. This type of project leader in product development has been termed 'heavyweight project manager' (or *shusa* in Japanese) in the literature (Clark and Fujimoto, 1991; Cusumano and Nobeoka, 1998). To sum up, the case illustrates the crucial role of tacit knowledge and product concepts in product development. It also teaches a lesson in how knowledge and value co-creation in the knowledge economy come about.

5.7 Maekawa Manufacturing

This case study illustrates how knowledge is co-created with customers through the co-creation of a joint *ba* between Maekawa and its customers. It also shows the importance of close interactions and long-term relationships with (key) customers and the sophisticated integration of product and service offerings.

5.7.1 Company information/background

Maekawa[35] Manufacturing's production of industrial freezers and associated systems began in 1924 and by now it has become one of the world's leading companies in industrial refrigeration systems. Since its inception, Maekawa has devoted itself to the accumulation of various forms of know-how (including elementary application and production technologies), and the creation of new markets by developing new products, focusing on customer needs in the food and thermal technology industries (von Krogh, Nonaka, and Ichijo, 1997). In 2000, Maekawa's share of the world's industrial refrigerator market was approximately 30 per cent and the firm plays an active and important role in the food and thermal control fields. Maekawa employs 3000 people (2250 in Japan and 750 overseas), with fifty-five overseas affiliated offices in twenty-eight countries (as of 2006). The freezing systems and heat-transfer technologies have been applied to move into various sectors in the energy industry. But Maekawa is not simply a machine manufacturer, compressor maker, energy consulting firm, or service company. Throughout its seventy-year history, it has greatly extended the spectrum of its activities to services and technologies in the fields of energy, food processing, and extremely low temperatures. Keeping in step with the times, Maekawa has developed into a full-service organization spanning a wide range of endeavours. Based on its unique management philosophy, trial and error research and development approach, customer-oriented business activities and, above all, its devotion to the spirit of challenge, Maekawa continues to grow and prosper, introducing new technologies, new services, and new ideals for its customers worldwide. Maekawa is an innovative company and has produced, for example, the world's largest compound two-stage helium compressor and a helium compressor for producing liquid hydrogen fuel (cf. also Peltokorpi and Tsuyuki, 2006).

5.7.2 Knowledge management initiatives/activities in general

There is basically no formal way of managing knowledge at Maekawa, nor is there a knowledge management department or knowledge manager. However, the unique project-based structure (see below) and their management philosophy strongly encourage the co-creation and sharing of knowledge. Research has cited Maekawa as a good example for enabler 2 – manage conversations – and enabler 4 – create the right context – (von Krogh, Ichijo, and Nonaka, 2000), as well as enabler 3 – mobilizing knowledge activists – (von Krogh, Nonaka, and Ichijo, 1997) of the knowledge enabler framework (Ichijo, 2006b). Peltokorpi and Tsuyuki (2006) identified the following set of knowledge governance mechanisms that Maekawa has created in order to promote knowledge processes: consensus-based hierarchy, shared HRM practices, performance measures, and output control. Discussing these in greater detail would go beyond the scope of this book, however. As for HRM practices, the frequent personnel rotation is certainly of particular importance for the sharing of tacit knowledge within and across the different units of Maekawa.

5.7.3 Knowledge-based marketing case: knowledge co-creation with customers

5.7.3.1 Organizational structure

Maekawa has rearranged its organizational structure several times, from a small entrepreneurial organization at the very beginning to a divisional, then a group-based organization, and finally from 1980 to the project-based *doppo* organization (see below and Figures 5.9 and 5.10) of today. This structure is really unique and can be described as a collective of many small 'independent companies'. Each of these independent companies consists of about twenty-five employees on average, and each of these small companies is established and classified by its product and/or market. Maekawa now consists of about eighty such corporations in Japan and about forty abroad. Each company either serves its local area or focuses on a specific market, for example, food, industrial freezers, or energy-related services. In a sense, they are like

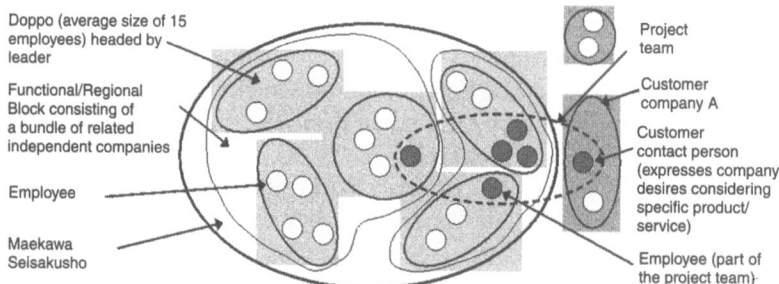

Figure 5.9 Maekawa's organizational structure with *doppos* and blocks and forming of project teams at Maekawa (from Peltokorpi and Tsuyuki, 2006: 41)

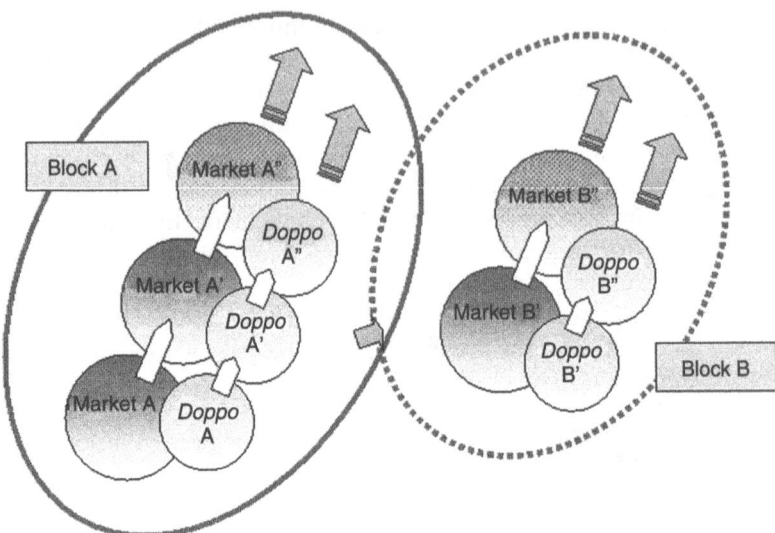

Figure 5.10 Markets, *doppos* and blocks (based on Maekawa, 2004: 128)

small independent venture businesses that constitute the Maekawa network. The purpose of this structure is to empower local organizations to cater to the specific needs of customers in niche markets as autonomous organizations. Each company is completely responsible for its own business and is self-sufficient, with a complete set of the functions it needs, from design to marketing. This so-called *doppo* structure is based on self-organization, meaning that units are relatively free to arrange themselves based on environmental stimuli. *Doppo* is short for *dokuritsu hojin*, meaning 'independent legal entity'. While *doppos* are financially independent, with their own accounting statements, the claim can be made that their prosperity and survival depend on their existence as integrated parts of Maekawa. The objective was to push decisions to where relevant knowledge and information reside and to increase environmental sensitivity, entrepreneurial action, and innovation by creating an organization in which the ideas of all employees can be leveraged.

Several *doppos* together can form *blocks*. Blocks are divided either by industries or by regions. As a result, they can comprise several different markets and the corresponding *doppos*. Figures 5.9 and 5.10 show Maekawa's organizational structure with *doppos* and blocks.

In principle, all entities are able to draw on each other's resources based on market needs. New *doppos* are established based on customer needs and are classified by product and market type, and in some cases, projects might become new *doppos*. Although employees identify themselves with their *doppo*, they share similar corporate values, vision, and beliefs. In this sense, Maekawa exists as a sort of community or collective entity. Indeed, although Maekawa's *doppos* are basically autonomous and self-sufficient, they are not isolated from each other. Some share the same office space and members from different *doppos* often spend time together to form informal relationships. Sometimes a new project or even a new *doppo* is created out of such relationships. In the end, Maekawa as a whole is a coherent organization with various parts organically interacting with each other.

5.7.3.2 Knowledge co-creation with customers

There are several cases that vividly highlight Maekawa's way of *ba* and knowledge co-creation with customers. Indeed, conversations with customers provide an important source of knowledge at Maekawa. At Maekawa, people do not ask for statistical data to read market trends. What they rely on are interactions with customers. It is not easy, however, to transform such interactions into a business plan (von Krogh, Ichijo, and Nonaka, 2000: 141–2). The independent corporations or task forces – *doppo* – are particularly important for Maekawa's efforts to develop new products. What makes this structural system unique is that the entities involved have more autonomy and authority than the standard task force (von Krogh, Ichijo, and Nonaka, 2000: 193). Customer needs and wants are largely tacit. When any member of a Maekawa *doppo* grasps such tacit knowledge or *ba*, there is in-depth communication and mutual understanding between the customer and the company, the kind that leads to development of an innovative product. As a result, Maekawa acquires tacit knowledge in a field through committed interactions with customers. But as such interactions cannot be accomplished overnight, employees must gain a wide spectrum of knowledge not only about customers' business but also all relevant social, economic, and environmental factors the company may be dealing with (von Krogh, Ichijo, and Nonaka, 2000: 197).

Maekawa focuses on making unique contributions to customers, which in turn requires solid understanding of customers' enabling contexts or *ba*, which can be achieved only by working closely with them so that knowledge will be created, captured, or capitalized on, resulting in a kind of 'co-innovation' with customers that depends on sharing tacit knowledge. A vivid example of this is the automated chicken deboning project (Toridas, see below) in which Maekawa engineers worked with employees at its customers' plants, resulting in such co-innovation (von Krogh, Ichijo, and Nonaka, 2000: 196).

The Toridas Project. The Toridas automated chicken deboning project has basically been going on for about twenty-five years now. In the early 1980s, in response to increasing labour costs in the industry, food-processing companies knew they had to increase the effectiveness and efficiency of their processes through automation. One idea that arose at Maekawa was to develop an automatic chicken leg deboning machine (named Toridas). But the process of chicken deboning is based on scarce tacit knowledge held by a small number of expert deboners. Maekawa promised to help the industry to increase automation, and selected companies involved in chicken deboning opened themselves to this supplier's engineers. These engineers focused for several months on chicken deboning, observing how the experts worked. After an eight-year development process, drawing heavily on the tacit knowledge they had gained of chicken deboning, as well as mechatronic and robotic technologies, Maekawa delivered a chicken deboning machine that truly fulfilled the expected returns. Toridas became one of the greatest commercial successes in its history. Yet now that Maekawa has satisfied their needs, its customers are reluctant to open themselves to other competitors, and this source of tacit knowledge is once again beyond the reach of those customers. In such a competitive situation, attempts at substitution become almost impossible (cf. von Krogh, Nonaka, and Ichijo, 1997: 481–2; von Krogh, Ichijo, and Nonaka, 2000: 76–7).

The Bread Factory Kaizen Project. This project started in 1997. Customer needs in the food (processing) industry had become more and more diverse and complicated, which meant that the suppliers had to provide more and more customized and flexible solutions. As a result, simply selling and installing parts or devices was no longer enough to successfully compete in the cost-driven market. For the bread factory *kaizen* project, four different Maekawa blocks joined forces to co-operate not only with each other but also with the customers, with whom business relations went back to the 1960s. The customer was a long established and renowned maker of European style bread in Japan and Maekawa had originally, among others, provided some cooling and freezing systems in the 1960s. But Maekawa wanted to enter more deeply into the bread-making market and help to improve the whole bread-making process. In order to do so, it was first necessary to really understand the process, which meant that Maekawa engineers had to learn how to make bread themselves as well. For this reason, Maekawa actually asked the customer to teach them how the whole process of bread making worked and thus got involved in the factory *kaizen* project through its own initiative. In the end, Maekawa and its customer were not only able to improve the process of making bread – achieving better flavour at the same time – but also to improve many other processes in the factory, including bulk production for wholesale. All of these efforts eventually led to a better quality product – tastier bread – at a lower production cost (cf. also Senoo, Akutsu, and Nonaka, 2001; Tsuyuki, 2001a).

5.7.4 Conclusion

This case study illustrates how knowledge is co-created with customers through the co-creation of a joint *ba* between Maekawa and its customers. It also shows the importance of close interactions and long-term relationships with (key) customers and the sophisticated integration of product and service offerings. As a matter of fact, both the co-creation of *ba* and building close interactions and long-term relationships with customers are closely related and interdependent. Although this is surely not a cause-and-effect relationship, the co-creation of *ba* – a shared context – requires and fosters the establishment of close, long-term relationships at the same time. Similarly, *ba* plays an essential role in this process of knowledge co-creation between different entities, be it members of a team at a single company or different entities of the business ecosystem. In the case of Maekawa, the transcendence of organizational boundaries occurs both within the company – that is, between the *doppos* – and between Maekawa and its customers. *Ba* nurtures mutual empathy and understanding and thus fosters relationships between people and companies, for example, between a company and its customers. In the end, this is a *conditio sine qua non* for knowledge and value co-creation in the network economy of today. Furthermore, the case once again underlines the importance of tacit knowledge and the need for direct interaction and face-to-face communication to co-create and share it. The co-creation of knowledge and value emerges from relationships and from the *ba* that is co-created between Maekawa and its customers through the long-term relationships. To sum up, the case illustrates that no company should be seen in isolation in the network economy any more. The co-creation of knowledge and value has become crucial in the business ecosystem (cf. also 4.2.2.2).

6
Knowledge-based Marketing: Results and Conclusion

As mentioned in Chapter 5.1, the case studies presented as vivid examples are not necessarily best practices, but rather explanatory practices, maybe also offering a first glimpse of 'next practices' (Prahalad and Ramaswamy, 2004a) that show how some global leading companies are on their way to successfully implementing and leveraging knowledge-based marketing for competitive advantage. This book has set out to explore current – and next – practices and historic circumstances of knowledge-based marketing through empirical research – expert interviews and participant observation – discussion and interaction with both scholars and practitioners, as well as a review and analysis of the relevant literature. Through action research, by developing a theoretical framework for knowledge-based marketing, and by explaining practical managerial implications, it aims to contribute to both marketing theory and practice.

In the process of conducting and analysing the expert interviews and the in-depth case studies of the empirical study, thoughts, ideas, and insights from the literature review became interwoven with each other and initiated the development of some sort of grounded theory, not unlike the grounded theory described by Glaser and Strauss (1967) and Strauss and Corbin (1990). Thus, a – still somewhat vague and multifaceted – picture of knowledge-based approaches to marketing has been emerging and reveals that there is a growing perception and recognition of the need for knowledge-based marketing but still only a few pioneer firms seem to be trying to meet this challenge.

The 'picture' – that is, the concept – of knowledge-based (approaches to) marketing was depicted and discussed in Chapter 4.2. The chapter also set out to provide definitions of the terms 'marketing knowledge' and 'knowledge-based marketing' and discussed important related issues. In order to illustrate the interweaving of theory and empirical insights, case studies of global companies that have successfully adopted knowledge-based approaches to marketing were portrayed in Chapter 5. The case studies were presented in the form of abbreviated vignettes, illustrating the essence of each informant company's knowledge-based approach. In fact, as they had been conducted as explanatory case studies, they were meant to highlight why and how the informant companies have adopted and implemented especially distinguished knowledge-based approaches to marketing.

This chapter will complement the 'picture' of knowledge-based marketing by discussing and analysing the case studies within the framework provided in Chapter 4.2. After the case analyses (section 6.1) a first taxonomy of knowledge-based marketing will be provided and discussed (section 6.2). Then, in section 6.3, the role of knowledge and knowledge-based approaches in the service-dominant logic for marketing will briefly be explored and a knowledge-dominant logic as a comprehensive framework of the knowledge-based 'picture' proposed.

Subsequently, section 6.4 presents the main conclusions of this book, section 6.5 briefly puts forward essential managerial implications, and section 6.6 looks at the limitations of this book and proposes implications for further research.

6.1 Case analyses

Despite the growing recognition of the need for knowledge-based approaches to marketing, there are only a few pioneer firms that are already taking or trying to take such an approach (see Kohlbacher, Holden, Glisby, and Numic, 2007; Kohlbacher, 2006). However, there are already some firms that are facing the challenge of an increasingly global business environment with fierce competition, taking up and mastering the challenge with the help of knowledge-based marketing. The focus of this book is specifically the co-creation of marketing knowledge within the firm – for example, between different units, functions, and so on – and with different entities within the business ecosystem of the firm.

Following the framework put forward in Chapter 4.2, specifically Figure 4.12 – which is reproduced here as Figure 6.1 – I will structure the discussion and the analysis of the six explanatory case studies.

6.1.1 Marketing knowledge co-creation within the firm

The first type of marketing knowledge co-creation takes place within the firm (cf. Figure 6.2). Knowledge is – or at least should be –

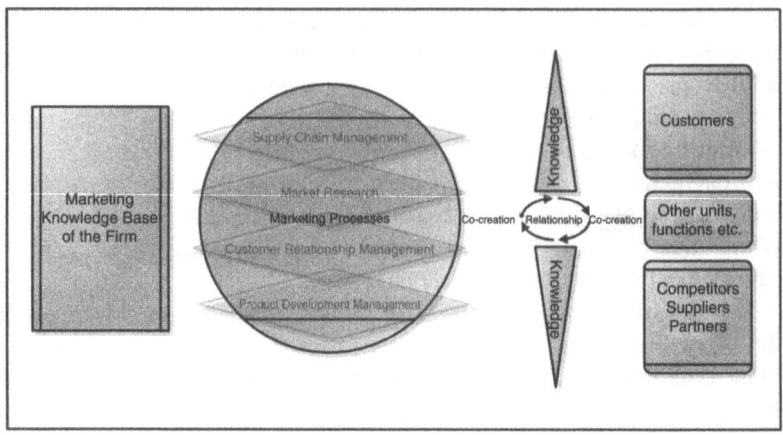

Figure 6.1 Knowledge-based marketing processes (integrated model II) (author's own illustration)

continuously (co-) created, shared, transferred, and so on, within a company. This involves different departments, business units, subsidiaries, and functions. The classic example would be the creation and transfer of knowledge between the headquarters and different subsidiaries of an MNC. Research on this is not uncommon and has already been cited in this book (for example, Schlegelmilch and Chini, 2003; Schlegelmilch and Penz, 2002; Simonin, 1999a). However, I assert

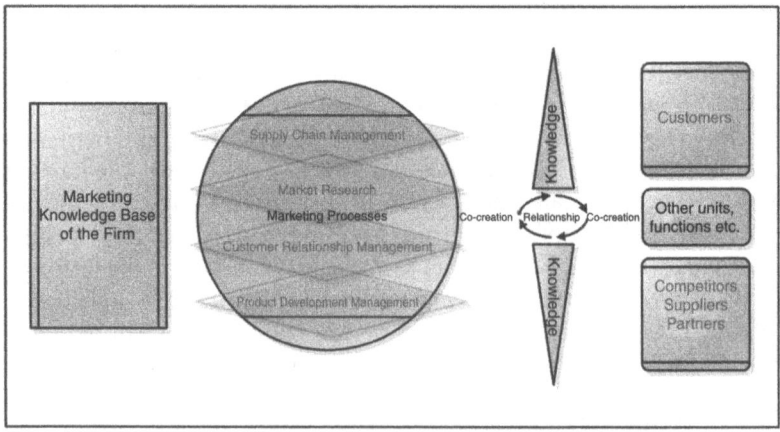

Figure 6.2 Marketing knowledge co-creation within the firm (author's own illustration)

that this alone would be too narrow a view of marketing knowledge and knowledge-based marketing processes within a firm, which is why I proposed holistic definitions of marketing knowledge and knowledge-based marketing in Chapter 4.2.

6.1.1.1 Hewlett-Packard Consulting & Integration (HP CI)

This case study illustrates how HP CI and specifically HP CI Japan leverage a particular form of CoP to create and share both tacit and explicit customer and other marketing knowledge. In Chapter 3.5 on CoPs, we saw that the sharing of expertise and the creation of new knowledge, often tacit in nature, are central tenets of a CoP's existence, whether they exist as a social gathering or technological network (Lave and Wenger, 1991). The sharing of tacit knowledge by and through CoPs is by means of storytelling, conversation, coaching, and apprenticeship provided by CoPs (Wenger, McDermott, and Snyder, 2002). As a matter of fact, the sharing of tacit knowledge – socialization – as well as its (partial) transformation into explicit knowledge – externalization – are at the heart of CoPs. This also seems to be in line with Nonaka's theory of knowledge-creation, and according to Plaskoff (2003: 179), '[c]ommunities provide an enabling context for knowledge creation'. Indeed, organization structures and systems that provide a context that co-ordinates and motivates action are critical elements of the overall knowledge organization (Wenger, McDermott, and Snyder, 2002). As a result, the concept of CoPs and the concept of *ba* share significant similarities, while they have important differences at the same time (cf. Nonaka and Toyama, 2003: 7; cf. also Nonaka, von Krogh, and Voelpel, 2006). HP CI Japan's learning communities are formally referred to as CoPs, but they obviously also provide the right context or *ba* for organizational knowledge creation. This is particularly important in the case of highly tacit knowledge about and from customers, as well as the experiences of the consultants and engineers in the field. This knowledge and these experiences provide very valuable feedback on customers, products, and services and help to further improve HP CI's value propositions.

Indeed, as shown above, critically important knowledge resides in the workplace, namely at the 'front lines' of the company (Yasumuro and Westney, 2001: 178) and most of this 'local' knowledge will be highly tacit (cf. also Yanow, 2004). In Japan these front lines are frequently referred to as *gemba*, which can be loosely translated as 'the actual spot or place', and according to Womack and Jones (2005b: 19) it is 'the Japanese word for the place in the office or factory where the real work is done'. Obviously, in HP CI's trade, a lot of knowledge is co-created at the *gemba*, the front lines of the company. Consultants and field engineers co-create new knowledge through interaction and teamwork among themselves, but most importantly, through interaction and co-operation with them they also co-create knowledge with customers. The learning communities then provide the *ba* to share this co-created knowledge and co-create at the same time new knowledge from the different experiences of their peers.

To sum up, the location of knowledge co-creation in the HP CI case is within the firm and the main co-creators of knowledge are HP CI's consultants and field engineers. The main type of marketing knowledge that is co-created and shared is customer knowledge (knowledge both from and about customers, as well as knowledge to support customers) and product and service knowledge. Finally, the marketing process in question here is CRM and the main method used for knowledge co-creation and sharing is CoPs (see Figure 6.3).

6.1.1.2 Schindler Elevator

This case study illustrates how Schindler Elevator took a knowledge-based approach to marketing in launching a new escalator product in Asian markets and effectively co-created and transferred both explicit and tacit knowledge between the Competence Centre Escalators in Vienna and Asia-Pacific headquarters in Hong Kong.

The Schindler case shows that even for a large MNC with its vast network of subsidiaries that are well connected by e-mail, intranet, databases, telephone, and video conferences, the sharing of tacit knowledge on a personal level is still a very reasonable or even indispensable approach. Schindler also realized that transferring knowledge one way

Location of knowledge co-creation	Main co-creators of knowledge	Main type of marketing knowledge	Marketing process	Knowledge co-creation method
Within the firm	Consultants, field engineers	Customer knowledge, product and service knowledge	CRM	CoP

Figure 6.3   Marketing knowledge co-creation at HP CI Japan

is not enough and that knowledge needs to flow freely both ways – in this case between the Competence Centre and subsidiaries and between master trainers and sales and marketing staff – to enable the creation of new knowledge that can lead to competitive advantage. This can be interpreted as the Swiss multinational's managing to combine both a codification and a personalization strategy for its marketing knowledge management. This is actually in contrast to Hansen, Nohria, and Tierney's (1999) recommendation to stick to either one of the two strategies (cf. also Umemoto, 2002). In fact, Schindler demonstrated how successfully combining a codification and personalization strategy as the approaches for sharing both tacit and explicit knowledge respectively, turned out to complement each other perfectly. However, it is also obvious that the personalization strategy is taken up only temporarily for major events like the market introduction of a new product (or a customer survey project in a different case), while the codification strategy of making all explicit knowledge and information available in a structured form via the intranet portal is long-term and continuous.

Moreover, Schindler also managed to combine two other important strategies, namely marketing exploitation and exploration strategies. As discussed in Chapter 4.2, both knowledge exploitation and knowledge exploration are indispensable for a company to increase its competitive advantage and 'market-oriented firms can gain important bottom-line benefits from pursuing high levels of both strategies in product development' (Kyriakopoulos and Moorman, 2004: 234). In the case of Schindler, knowledge exploitation took place when existing knowledge was shared commonly in the training workshops and on the intranet. On the other hand, knowledge exploration was triggered by the exhaustive discussions and by the feedback from the subsidiaries' staff and led to the creation of new knowledge for future product development.

Furthermore, in the case of Schindler, a new knowledge sharing and transfer hub was created at the Asia-Pacific headquarters in Hong Kong by temporarily dispatching experts from the Competence Centre in Vienna. They not only transferred their own expert knowledge as knowledge activists but also served as knowledge brokers[36] to enable and arrange the flow of knowledge between different subsidiaries in the Asia-Pacific region and headquarters. Using this approach, Schindler was able to facilitate value creation within the firm, a task for which marketing functions are highly dependent on knowledge transfer within the organization (Schlegelmilch and Chini, 2003). Besides, as will be recalled '[m]arket knowledge is not fully captured in a usable form until the lessons and insights are transferred beyond those who gained the experience' (Day, 1994b: 23) and thus, by enabling and fostering the multilateral transfer and sharing of knowledge, Schindler made sure that its market and marketing knowledge could be captured and added to its corporate knowledge base.

To sum up, the location of marketing knowledge co-creation in the Schindler case was within the firm, but between different subsidiaries, specifically between the Competence Centre Escalators in Vienna, Asia-Pacific headquarters in Hong Kong, and local subsidiaries in Asia. The main co-creators of knowledge were the product managers (called product line managers at Schindler), marketing and salespeople, as well as field engineers, with the main type of knowledge being product and service knowledge. The marketing process involved was market introduction/product launch, which can be seen as located somewhere between PDM and CRM. Finally, the main method for knowledge co-creation was the dispatching of experts and their interaction with local staff at work in general, as well as during marketing workshops in particular (see Figure 6.4).

6.1.1.3 Siemens

This case study presents a recently launched company-wide cross-selling and marketing knowledge sharing initiative in Siemens and illustrates how cross-functional, cross-divisional, and cross-regional collaboration is leveraged for value and knowledge co-creation, resulting in superior value propositions to customers.

Location of knowledge co-creation	Main co-creators of knowledge	Main type of marketing knowledge	Marketing process	Knowledge co-creation method
Within the firm; between different subsidiaries	Product managers, marketing managers, sales managers, sales people and field engineers	Product and service knowledge; marketing knowledge in general	PDM, CRM	Dispatching of experts, workshops

Figure 6.4 Marketing knowledge co-creation at Schindler Elevator

In order to be successful, the Siemens One initiative had to adopt a knowledge-based approach to marketing. Therefore, CRM activities, account planning, sector development, and co-creation with customers are essential pillars that back up the Siemens Customer Focus programme.[37] Senn (2006), for example, has discussed Siemens' approach to CRM including the Top Executive Relationship Programme whose purpose is to foster the relationship with strategic key accounts through personal interaction at the top management level. Indeed, the starting point and centre of Siemens' approach to knowledge-based marketing is its 'customer focus programme' and the resulting customer-focused organizational structure. As mentioned in Chapter 5.4.4, Siemens One is a very comprehensive case of knowledge-based marketing in action as it not only spans the boundaries of different business divisions within the MNC but also transcends the boundaries of the firm and involves customers, partners, and so on.

Writing about fifteen years ago, Achrol (1991: 80) already argued that 'the firm of the future will need to be very permeable across its departments' and that 'to be effective, individuals will need to be skilled in lateral relations – in cooperating and winning cooperation, in assembling and reorganizing the right groups of talent around problems and solutions'. In fact, there is a cost to this functional specialization, which is 'the development of silos of information where information is isolated and is principally contained in those same functional areas' (Grieves, 2006: 66). As a result, cross-functional teams or multifunctional teams and the resulting knowledge integrations (Clark and Fujimoto, 1990; Natter, Mild, Feuerstein, Dorffner, and Taudes, 2001: 1030) have become core issues – specifically in product development – and also led to the visions of 'One Company' (Grieves, 2006: 235), as can be seen in the case of the Siemens One initiative. In fact, the rise of teamwork together with the loss in importance of functional boundaries has been attributed to the need to create new knowledge within the firm (Sinkula, 1994; Slater and Narver, 1995), to share information across functional boundaries (Jaworski and Kohli, 1993; Narver and Slater, 1990), and last but not least to respond more rapidly to changes in the market (Achrol, 1997; Griffin, 1997), all of which are important steps in the evolution towards a customer-focused marketing organization (Homburg, Workman, and Jensen, 2000).

An important initiative to foster and drive the customer focus is the implementation of the Siemens One organization. One of its main goals is to enable and harness cross-group – that is, between different business units at Siemens – and cross-regional (or cross-country) collaboration and to act as one Siemens in the market, namely to just show one face to the customer. In fact, the focus of the Siemens One approach is the establishment of a strategic relationship with certain clients and the provision of solutions for selected market segments, such as airports, hospitals, and stadiums. These activities are supported by the implementation or reinforcement of KAM (cf. also Homburg, Workman, and Jensen, 2000; Senn, 2006; Wengler, Ehret, and Saab, 2006) and sector approaches. Besides, a persistent service-orientation and added-value by bundling and integration of its products and services also help to drive and promote Siemens' endeavours to become a customer-focused marketing organization. Leveraging cross-group and cross-regional collaboration and synergies, providing value-added by offering bundled or integrated solutions is not only believed to attract and win new customers but also to enable cross-selling to existing customers as well.

To sum up, the location of marketing knowledge co-creation in the Siemens case is basically within the firm, between different departments, business units, and subsidiaries. The main co-creators of knowledge are product managers, marketing managers, and salespeople, and the main type of marketing knowledge is customer knowledge, but also product and service knowledge. The main marketing process involved is CRM and the main methods of knowledge co-creation are cross-functional and cross-divisional workshops (for example, Account Planning Process Plus, APP+) and the cross-functional and cross-divisional teams and their collaboration in general (Figure 6.5).

Location of knowledge co-creation	Main co-creators of knowledge	Main type of marketing knowledge	Marketing process	Knowledge co-creation method
Within the firm; between different business units and subsidiaries	Product managers, marketing managers, sales managers, sales people	Customer knowledge, product and service knowledge	CRM	Workshops, cross-divisional and cross-functional teams

Figure 6.5 Marketing knowledge co-creation at Siemens

6.1.2 Marketing knowledge co-creation with competitors, suppliers, and partners

The second type of marketing knowledge co-creation takes place between the firm and certain stakeholders in its business ecosystem (cf. Figure 6.6). Knowledge is not only – or at least should not be – continuously (co-)created, shared, transferred, and so on, but also with other entities outside a company. This involves, for example, competitors, suppliers, and partners (customers will be dealt with below, 6.1.3). The explanatory case below shows how Toyota strategically co-creates knowledge with a competitor, but Toyota is also known for the knowledge creation and sharing within its supplier system (cf. also 6.2).

6.1.2.1 *Toyota Motor Corporation/Toyota Peugeot Citroën Automobile Czech (TPCA)*

This case study illustrates how Toyota leverages marketing knowledge through a JV with a competitor in order to expand its business in Europe and crack the East European market and also co-creates new knowledge with the alliance partner. It depicts the case of Toyota's effort to expand and reinforce its presence and sales in Europe, especially in the emerging economies of Eastern Europe. Here, Toyota's approach to exploitation and exploration of marketing knowledge was implemented by joining forces with a major player in the European automotive industry.

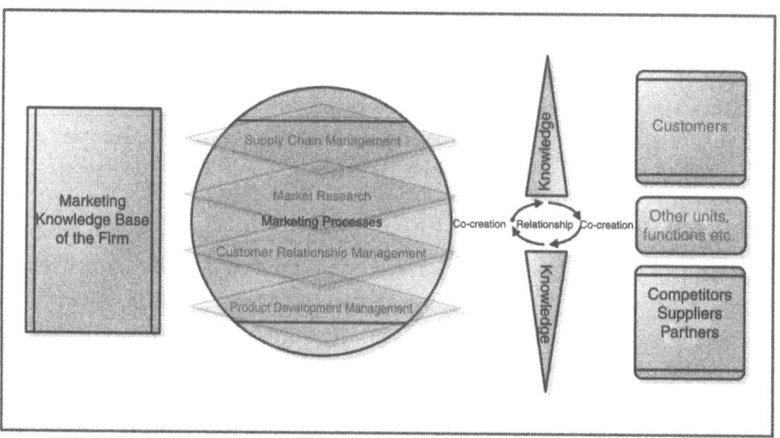

Figure 6.6 Marketing knowledge co-creation with competitors, suppliers, and partners (author's own illustration)

Recently, Toyota's knowledge creation in automotive development has changed from creating new knowledge in Japan and transferring it from headquarters to subsidiaries and affiliations around the globe to a focus on creating knowledge in foreign markets by local staff (cf., for example, Ichijo and Kohlbacher, 2006b, 2007). With its new strategy of 'learn local, act global' for international business development, Toyota proved successful in tapping rich local knowledge bases, thus ensuring its competitive edge and global lead in the automotive industry. TPCA has provided a fast, cost-efficient response to market demand through the sharing of expertise and experience. As shown above, leveraging synergies and fostering mutual knowledge sharing and creation between the two partners are two of the most important goals and merits of this strategic alliance. As a consequence, it is probably safe to say that Toyota has made the leap from simply being a global projector to a truly metanational company and this going beyond transnational strategy has been identified as especially crucial for entering emerging markets (London and Hart, 2004).

The collaboration with PSA is expected to result in mutual learning and other advantages for Toyota's European operation and its strategy of using external, local resources (Ando, 2005). This attempt by organizations to realize their objectives through co-operation with other organizations rather than in competition with them is called 'co-operative strategy' (Child, Faulkner, and Tallman, 2005) and as Simonin (1999a: 485–6) notes, as we enter 'a marketing era characterized by globalizing marketplaces and by the necessity to both rapidly develop and effectively protect core competencies, international strategic alliances are becoming even more critical for companies to sustain competitive advantage'. As for the TPCA case, Toyota considers this collaboration as one of its efforts to meet consumer demand for low-cost, fuel-efficient, and environment-friendly vehicles and believes that co-operating with PSA will provide a capable response to the expanding small passenger car market.

Furthermore, by consistently following its 'learn local, act global' strategy, Toyota is feeding back newly created and acquired knowledge to its headquarters and spreading it also to other subunits. Indeed, in Toyota's continuous learning system, '[t]ough analysis, reflection, and communication of lessons learned are central to improvement as is the discipline to standardize the best-known practices' (Liker, 2004: xvi). As shown above, critically important knowledge resides in the workplace, namely at the 'front lines' of the company (Yasumuro and Westney, 2001: 178) and most of this 'local' knowledge will be highly tacit (cf. also Yanow, 2004). Toyota's principle of *genchi genbutsu* and the role of *gemba* are critical in this context. According to Liker (2004: 224, original emphasis), the 'first step of any problem-solving process, development of a new product, or evaluation of an associate's performance is grasping the actual situation, which requires "going to *gemba*"'. In the case of TPCA, this kind of critically important knowledge will be created at the front lines and *gembas* of the JV, through the interaction of the partners, but also through interaction with suppliers and customers. The process of co-creation and transfer of knowledge between Toyota and PSA expatriates, as well as with local staff, will be crucial for the success of the JV. Indeed, the way Toyota and PSA divided up roles and responsibilities shows that TPCA is a true co-operation and really managed jointly by both partners. This way, interaction and direct communication between staff dispatched from both parents are high and thus may be assumed to offer superior opportunities for knowledge (co-)creation and sharing, as well as for mutual learning.

As the IMV project and the TPCA case have illustrated, Toyota can be seen as a firm with a strong market orientation, possessing 'the basis for rapid adaptation to customers' manifest and latent needs, which may translate into superior new product success, profitability, market share, and, perhaps, sustainable competitive advantage' (Baker and Sinkula, 2005: 483). Hence, we can conclude that Toyota has successfully implemented knowledge-based marketing.

To sum up, the main location of marketing knowledge co-creation in the Toyota case was between competitors and the main co-creators of knowledge were product, marketing, and other middle and top managers. The main type of knowledge co-created was supplier, product, and production knowledge. The main marketing processes involved were PDM, SCM, but also partly market research and CRM. Finally, the knowledge co-creation method was the formation of a JV, namely a strategic alliance and the direct interaction between the two partners (Figure 6.7).

6.1.3 Marketing knowledge co-creation with customers

Finally, the third type of marketing knowledge co-creation takes place between the firm and a particularly essential stakeholder in its business ecosystem (cf. Figure 6.8). This stakeholder is of course the customer. The two explanatory case studies – Mazda Motor Corporation and Maekawa Manufacturing – tried to illustrate this in greater detail.

Location of knowledge co-creation	Main co-creators of knowledge	Main type of marketing knowledge	Marketing process	Knowledge co-creation method
Between competitors	Product managers, marketing managers, and other middle and top managers	Supplier knowledge, product knowledge, production knowledge, marketing knowledge in general	PDM, SCM, market research CRM	JV/strategic alliance

Figure 6.7   Marketing knowledge co-creation at Toyota/TPCA

6.1.3.1 Mazda Motor Corporation

This case study illustrates how an essential product concept and related tacit marketing knowledge were passed on and refined over three product generations. It also shows that the experience of co-creation between the product and the customers is crucial and that empathic design to tacitly understand customer needs is vital.

As we have seen in the case study, developing, disseminating, and implementing a unique concept is an essential step in product develop-

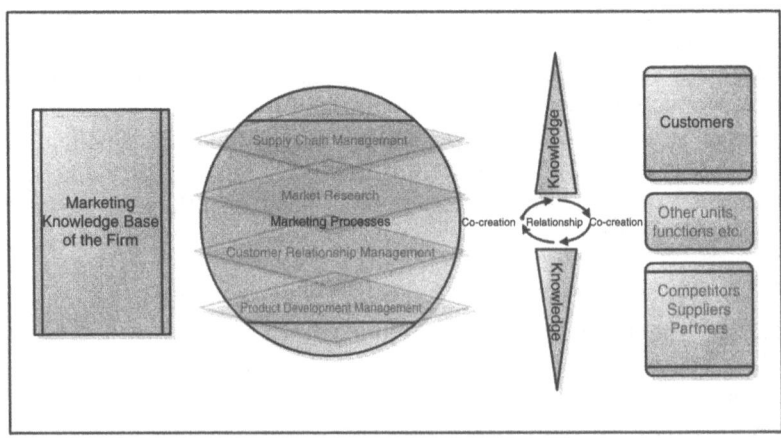

Figure 6.8   Marketing knowledge co-creation with customers (author's own illustration)

ment. But these concepts usually tend to be highly tacit and as such difficult to transfer to others. Indeed, especially in the concept development stage, it is critical to articulate images rooted in tacit knowledge and meaningful information arises as a result of the conversion of tacit knowledge into articulable knowledge (Nonaka, 1990a). Explaining the process of externalization of tacit knowledge into explicit knowledge, Nonaka and Takeuchi (1995: 64–5) maintain that a 'frequently used method to create a concept is to combine deduction and induction' and highlight the example of Mazda, which combined these two reasoning methods when it developed the new RX-7 concept. In fact, as Nonaka and Toyama (2003) argue, abduction or retroduction might be even more effective than induction or deduction to make a hidden concept or mechanism explicit out of accumulated tacit knowledge. In the roadster case, the use of symbolic language and the metaphorical concept of *Jinba Ittai* proved essential for the success of the car for generations.

Metaphors – like *Jinba Ittai* and 'lots of fun' – can also be used for identifying and learning about customer needs, as shown in Chapter 4.1.3.2. Using the fishbone chart and discussing the concept on the basis of the interpretations from the concept catalogue were helpful measures for an externalization of the tacit concept. As discussed in Chapters 4.1.3.2 and 4.1.3.4, this way of capturing customer needs and translating them into a product concept has been termed 'empathic design'. The product development team was indeed able to capture customer needs and translate them into a successful product concept because of their capacity to leverage even tacit customer needs and knowledge and achieve a high level of experience co-creation between the customers and the product. Obviously, Hirai and Kijima are real masters of empathic design. *Kansei* engineering is an especially useful tool in this respect. For example, Mazda tested over 150 potential tunings for the roadster's exhaust system, which was done to match the sound of the MX-5 to the customer's perceived idea of what a roadster should sound like. Indeed, it is exactly these 'unexpected touches of quality that go beyond the obvious' and meticulousness (Hodock, 1990: 7; Kotler, Jatusripitak, and Maesincee, 1997: 352), as well as the role of sensation and aesthetic experiences which are important for knowledge-based marketing and innovation.

Transferring and refining – and thus *recreating* – the concept of *Jinba Ittai*, as well as the technique of *Kansei* engineering was fundamental to the success of the Mazda roadster. In fact, as Herbig and Jacobs (1996: 66–7) have pointed out, 'Japanese innovation refers to the application, the refinement of an idea', and '[i]deas from many people are gathered, assimilated, and squeezed into a new product or solution'. This is also exactly what happened in the case study. This is a very important point as it shows that time and experience play an important role. As Cusumano and Nobeoka (1998: 176) put it: 'Engineers need to learn how to do this kind of design work through experience, and what they need to do may vary widely from project to project. It is difficult to write down or codify this type of knowledge.'

The development of the first-generation roadster can certainly be seen as a radical innovation. At that time, nobody saw a market and business opportunity for a lightweight sports car. But the Mazda roadster turned out to be an outstanding success and it created and opened up a whole new market for Mazda. In the terminology of Kim and Mauborgne (2005; 1999) the innovation can also be interpreted as a 'value innovation'. Instead of trying to outperform competitors through better quality or new technological features, and so on, value innovation 'makes the competition irrelevant by offering fundamentally new and superior buyer value in existing markets and by enabling a quantum leap in buyer value to create new markets' (Kim and Mauborgne, 1999: 43). With the development of the roadster, Mazda managed to 'break out of the competitive and imitative trap' (ibid.).

Finally, according to Prahalad and Ramaswamy (2003), the value of products or services is in the co-creation experience that stems from the customer's interaction with the product and/or the firm (cf. also Prahalad and Ramaswamy, 2004a). In the case of the roadster, the value for the customer is created when he/she actually drives the car and feels and experiences both 'lots of fun' and *Jinba Ittai*.[38] Indeed, innovations in the knowledge economy can also be seen as co-creations of experiences with customers (for example, Prahalad and Ramaswamy, 2003, 2004a).

To sum up, the location of marketing knowledge co-creation in the Mazda case was within the firm – specifically in the PDM process – but also between the firm and its customers. The main co-creators of knowledge were product managers and managers and engineers involved in product development, with the main type of marketing knowledge co-created being customer and product knowledge – specifically product concepts. The main marketing processes involved were PDM and market research, in a broader sense maybe even CRM. The main method of knowledge co-creation can be summarized as empathic design (Figure 6.9).

6.1.3.2 Maekawa Manufacturing

This case study illustrates how knowledge is co-created with customers through the co-creation of a joint *ba* between Maekawa and its cus-

Location of knowledge co-creation	Main co-creators of knowledge	Main type of marketing knowledge	Marketing process	Knowledge co-creation method
Within firm, with customers	Product managers, product development engineers	Customer knowledge, product knowledge	PDM, market research, CRM	Empathic design

Figure 6.9   Marketing knowledge co-creation at Mazda

tomers. It also shows the importance of close interactions and long-term relationships with (key) customers and the sophisticated integration of product and service offerings.

Maekawa Manufacturing Ltd. – famous for its decentralized structures and project-based management approach (for example, Peltokorpi and Tsuyuki, 2006; Senoo, Akutsu, and Nonaka, 2001) – found that producing and selling industrial parts is no longer enough. Through co-creation of common contexts and knowledge with its customers it was able to combine its products with its process knowledge to offer an integrated service, including consulting, which means that Maekawa has transformed itself from a supplier of physical products and parts to a provider of comprehensive solutions. However, they do not simply offer pre-defined processes and manufacturing models, but actively co-create the solutions together with their customers (cf. also Maekawa, 2004; Tsuyuki, 2001a), an achievement that helped them to escape the red oceans of cut-throat competition and create new market space (blue ocean), yet 'untainted by competition' (Kim and Mauborgne, 2004: 77; cf. also Kim and Mauborgne, 2005). Maekawa indeed has an explicit philosophy of *mu-kyoso*, 'no-competition', as they believe that it is necessary 'to go one's own way' and 'ignore the competition'. This is also manifested in some of Maekawa's publications, most prominently 'From Competition to Co-creation' (Shimizu and Maekawa, 1998). This is consistent with Prahalad and Ramaswamy's (2004a: 12, original emphasis) argument that the competition needs to be made 'irrelevant' and that the future of competition 'lies in an altogether new approach to value creation, based on *an individual-centered co-creation of value between consumers and companies*'.

Engineers at Maekawa's independent corporations – *doppo* – are therefore strongly encouraged to develop an ability to talk with customers.

This is considered an indispensable aspect of their participation in that enabling context. Engineers who can recognize the technical needs of customers by observing their production lines with them are trusted more than those focusing only on technical specifications (cf. also von Krogh, Ichijo, and Nonaka, 2000: 198). These direct interactions and the trust are absolutely necessary in co-creating and nurturing a shared context or *ba* between Maekawa and its customers. Moreover, as Maekawa aims at building and sustaining relationships with its customers long-term rather than merely seeking short-term financial profit, these shared contexts with customers grow and are fostered over time. Establishing long-term relationships, mutual trust, and joint *bas* enables effective co-operation and teamwork between Maekawa and its customers and at the same time harnesses the co-creation and sharing of knowledge. As a matter of fact, both the co-creation of *ba* and building close interactions and long-term relationships with customers are closely related and interdependent. Although this is not a cause-and-effect relationship, the co-creation of *ba* – a shared context – requires and fosters the establishment of close, long-term relationships at the same time. Here, the co-creation of *ba* means that *ba* is not only created and established within one organization, but actively and jointly created together with and between organizations, specifically Maekawa and its customers.

As mentioned in Chapter 3.4.1, Nonaka, Toyama, and Nagata (2000) distinguish between four different types of *ba*. As the co-creation of knowledge and value between Maekawa and its customers is deeply rooted in tacit knowledge and face-to-face communication and interaction, the types of *ba* in question are obviously 'originating *ba*' and 'dialoguing *ba*', which are defined by individual or collective and face-to-face interactions respectively. However, even though the distinction of the four types of *ba* is theoretically and analytically possible, in reality the distinction is difficult and there will rather be a mix of different *bas* and the *bas* will constantly switch between the different types in the course of time.

This way of collaborating with customers to create superior service and innovation leads at the same time to the co-creation of a vast amount of – often very tacit – product and service knowledge but also knowledge from and about customers, as well as obviously knowledge to support customers. This knowledge is utilized and refined over years in long-term relationships with a customer, but it can also often be applied to projects with other customers as well. Besides, collaborating with their customers means also to think beyond them and also about

Location of knowledge co-creation	Main co-creators of knowledge	Main type of marketing knowledge	Marketing process	Knowledge co-creation method
With customers	Field engineers	Customer knowledge, product and service knowledge	CRM, PDM	Co-creation of *ba*

Figure 6.10 Marketing knowledge co-creation at Maekawa

the customers' customers. As a result, the co-created customer knowledge also includes knowledge about, from, and to support customers' customers. Indeed, Maekawa is not only concerned about the quality of its own products and services but also about that of its customers. This is a very special and important value proposition.

To sum up, the main location of marketing knowledge co-creation in the Maekawa case was with customers and the main co-creators of knowledge are field engineers. The main type of marketing knowledge co-created is customer, product, and service knowledge. The marketing processes involved are CRM and PDM and the main method of knowledge co-creation is the co-creation of a joint *ba* (Figure 6.10).

6.2 Summary: a taxonomy of knowledge-based marketing

As mentioned at the beginning of Chapter 5, the six explanatory case studies of knowledge-based approaches to marketing that are presented and discussed in this book show how things could and can be done differently in two senses. First, the six companies have consciously opted to pursue knowledge-based approaches to marketing and thus act differently from many of their competitors. And second, the approaches of the six companies – despite some similarities – also differ from each other and thus offer insights on different knowledge-based strategies. This can be summarized as follows (see also Figure 6.11).

(1) The HP CI case vividly illustrates the significance of tacit knowledge from *gemba* or the front line of the company. Knowledge is shared and co-created in a shared context or *ba*, in this case the so-called learning communities. It highlights also the importance

of knowledge from, about, and to support customers as a *sine qua non* to provide superior service and value propositions.

(2) The Schindler case illustrates the exploitation and exploration of marketing knowledge by combining both a personalization and a codification strategy in a new product market introduction project. The focus here is on the collaboration and effective communication – but also the co-creation – of both tacit and explicit marketing knowledge between headquarters (that is, the Competence Centre Escalators in Vienna and Asia Pacific headquarters in Hong Kong) and subsidiaries (for example, Japan, China, other Asian countries, and so on) and between subsidiaries and thus goes far beyond what Dixon (2000) has termed 'strategic transfer' (cf. also Chapter 3.6).

(3) In a similar vein, the Siemens case shows how important (marketing) information and knowledge is collected, co-created, shared, and applied both across business units and also across subsidiaries or regions/countries. It takes the approach a step further through its customer focus and KAM approach, which includes direct interaction and co-creation of knowledge with customers. As mentioned above, it is therefore a very comprehensive case of knowledge-based marketing which actually goes far beyond the focus on marketing knowledge co-creation within the firm presented in this book.

(4) The Toyota case has illustrated the importance of the co-creation of knowledge with competitors, especially when entering new and as yet rather unexplored markets. Here mutual learning and knowledge co-creation not only boost the performance of the strategic alliance, but lessons learned can also be transferred and implemented at the respective headquarters and other subsidiaries.

(5) The Mazda case has depicted the necessity to empathize with customers and to directly 'experience' their needs. This gained tacit knowledge then needs to be translated or externalized into explicit knowledge and product concepts, for example, with the help of metaphors. Thus, this case also shows the importance of language for knowledge creation, as well as the critical issue of re-creating and refining tacit knowledge over time. Last but not least, it also highlights the significance of value co-creation with customers.

(6) The Maekawa case shows the importance of co-creating shared contexts or *bas* with customers and of establishing and nurturing long-term relationships with them. Consequently, the mutual understanding and co-creation of tacit knowledge is crucial for

‌

‌‌

‌‌‍

‌‌‍

Informant company	Approach	Main co-creator(s)	Classification
HP CI	Intra-firm knowledge co-creation and sharing: CoPs	Local staff/front-line employees	Intra-firm co-creation
Schindler Elevator	Intra-firm knowledge creation and sharing: inter-subsidiary	Local staff	Intra-firm co-creation
Siemens	Intra-firm knowledge creation and sharing: inter-divisional and inter-regional	Local staff (in different countries and different business dividions)	Intra-firm co-creation
Toyota/TPCA	Inter-organizational knowledge creation and sharing	Competitor	Competitor co-creation
Mazda	Intra-firm knowledge creation and sharing; Value co-creation with customers	Local staff (product development team members)	Intra-firm co-creation; customer co-creation
Maekawa	Inter-organizational knowledge creation and sharing	Local staff/front-line employees	Customer co-creation

Figure 6.11 Summary: a taxonomy of knowledge-based marketing

creating superior solutions and value jointly in the business eco-system of a firm.

Specifically Chapters 4, 5, and 6 were meant to answer the research question posed in Chapter 2.2: *What is knowledge-based marketing and which types and patterns of marketing knowledge co-creation within MNCs can be identified?*

Indeed, Chapter 4 has provided both a definition and a conceptual framework of knowledge-based marketing and different types and patterns of marketing knowledge co-creation, and Chapters 5 and 6 have illustrated and discussed this with the help of real-life examples from the corporate world. The summary of this chapter now provides a taxonomy of knowledge-based marketing as it was identified in the empirical research project underlying this book. Similar to Dixon's (2000: 143) conclusion that '(1) there are many, very different ways to transfer knowledge, and (2) knowledge is transferred most effectively when the transfer process "fits" the knowledge being transferred' (see also Chapter 3.6), one answer to the research question could be that there are many, very different ways – types and patterns – of marketing knowledge co-creation within MNCs and marketing knowledge is co-created most effectively when the co-creation process – or the way of co-creation – fits the knowledge being transferred and the circumstances of the company. This book has identified three different basic patterns – co-creation within the MNC, co-creation with competitors, suppliers, and partners, and co-creation with customers – and has provided six different exemplifying case studies for these. The important results here are not necessarily the ways of co-creation and how they can be compared, but rather the fact that co-creation of knowledge and value in the business ecosystem is crucially important, with the identified cases serving as guiding real-life examples of how knowledge-based firms do it in practice. Concerning the two sub-questions *What is marketing knowledge?* and *What is its role in marketing and how is it created and managed?* the former has been answered in Chapter 4.1.2 and particularly 4.2.1, and further illustrated by the case studies. The latter has equally been explored and answered in Chapter 4.2 – particularly sections 4.2.2 and 4.2.3 – and through the case studies and their discussion. The four conclusions in section 6.4 below also highlight the main findings and results from this research project.

As discussed in the Appendix, the case studies were sampled and selected as critical cases that show different types of knowledge-based approaches to marketing. But obviously there seem to exist a large variety of such approaches, not just six. Besides, the cases or types/ patterns are by no means mutually exclusive of each other. It was just for the sake of structure and clarity that the six explanatory cases were meant to exemplify one single, typical approach. As also shown in Chapter 4.2, firms should engage in marketing knowledge co-creation not only within their own boundaries but also with other entities within the business ecosystem. The informant companies in this research project do so, too. Siemens Japan, for example, engages in co-creation of knowledge with business partners (for example, Marubeni) and competitors or other firms through JVs (for example, Yaskawa, NEC, Fujitsu, and so on).

Moreover, learning in the automotive industry has frequently been a research topic by itself (cf., for example, West, 2000) and cases of alliances are abundant as well (cf., for example, Doz and Hamel, 1998; Ghosn, 2002; Inkpen and Ramaswamy, 2006). Especially the JV between Toyota and GM – New United Motor Manufacturing, NUMMI – has already become legendary and has repeatedly been discussed (cf., for example, Badaracco, 1991; Inkpen, 2005; O'Reilly III and Pfeffer, 2000, to name just a few). In Chapter 4.1.3.1 we looked at knowledge co-creating with suppliers. In the automotive sector, the *keiretsu* structuring of supplier relations 'historically enabled Japanese auto assemblers to remain lean and flexible while enjoying a level of control over supply akin to that of vertical integration' and the '[h]igh trust, long-term cooperation between assemblers and their suppliers has made possible reductions in new model development time in the Japanese auto industry' (Ahmadjian and Lincoln, 2001: 683). Dyer and Nobeoka (2000: 346) further note that 'Toyota and other leading Japanese automakers (notably Honda) have developed bilateral and multilateral knowledge-sharing routines with suppliers that result in superior interorganizational or network-level learning.' This seems to be consistent with Fujimoto's (1999) view that Toyota's competitive edge comes in part from its ability to work with a set of independent suppliers to create knowledge (cf. also Dyer and Hatch, 2004; Dyer and Nobeoka, 2000; Evans and Wolf, 2005; Liker, 2004; Liker and Choi, 2004). Dyer and Hatch (2004: 58, original emphasis), for example, found that 'the company has developed an infrastructure and a variety of *inter*organizational processes that facilitate the transfer of both explicit and tacit knowledge within its supplier network', and Wenger, McDermott, and Snyden (2002) also point out Toyota's communities in business-to-business clusters, such as their supplier networks. Yet, even though Dyer and Nobeoka (2000: 347) contend that 'Toyota's "network" appears to be highly effective at facilitating interfirm knowledge transfers and may be a model for the future', they have to admit that 'at present the collaborative process used by Toyota to facilitate these transfers is somewhat of a black box'.

However, creating knowledge just for the sake of creating it will most likely not lead to sustainable competitive advantage. Knowledge-based management and the process of knowledge creation must aim at generating innovation and as a result higher value propositions to the firm's environment, specifically its customers but also all other stakeholders. Besides, particularly in the case of JV and alliances, even though '[t]he formation of an alliance is an acknowledgement that an alliance partner has useful knowledge' (Inkpen, 1998: 71), this does not necessarily lead to mutually beneficial learning and knowledge (co-) creation. Indeed, Hamel's (1991) 'learning race' has become well known and so has the distinction between 'learning alliances' and 'cospecialization alliances' (Mowery, Oxley, and Silverman, 2002). These two learning situations within an alliance have also been termed 'collaborative learning' – based on an underlying spirit of collaboration between the partners – and 'competitive learning' – based on an underlying attitude of competition between them (Child, Faulkner, and Tallman, 2005). The same is basically also true for knowledge cocreation within a firm or with other entities in the business ecosystem.

6.3   Toward a knowledge-dominant logic for marketing

> Neither art nor science stands still in representing our visible and invisible worlds. Marketing, as both art and science, can't stand still either. (Zaltman, 2003: ix)

> The three value drivers – customer value, core competencies, and collaborative networks – are leading to a new marketing paradigm. (Kotler, Jain, and Maesincee, 2002: 25)

'Marketing theory must reinvent itself and be refined, redefined, generated, and regenerated – or it will inevitably degenerate' and '[m]uch of marketing knowledge resists time and change and should be retained', while 'much is obsolete or was never up to par and should be dumped, and much is not yet discovered' (Gummesson, 2005: 317). According to Webster (2005: 125), '[f]undamentally new paradigms of marketing management are being offered that shift the core focus of the field from firms to customers, from products to services and benefits, from transactions to relationships, from manufacturing to the co-creation of value with business partners and customers, and from physical resources and labor to knowledge resources and the firm's position in the value chain'. Vargo and Lusch (2004) argue that an evolution is under way toward a new dominant logic for marketing, one in which service proposition rather than goods is fundamental to economic exchange (cf. also Lovelock and Gummesson, 2004). This worldview has become known as 'service-dominant logic' and posits that 'marketing has moved from a goods-dominant view, in which tangible output and discrete transactions were central, to a service-dominant view, in which intangibility, exchange processes, and relationships are central' (Vargo and Lusch, 2004: 2). According to Leonard (1998: xiii), products are physical manifestations of knowledge, and their worth largely, if not entirely, depends on the value of the knowledge they embody. Zaltman (2003: x) contends that in 'this "knowledge-explosion epoch" ... the limitations of our current marketing paradigm loom ever larger'.

In proposing a new dominant logic for marketing, Vargo and Lusch (2004: 1–2) posit that 'marketing has shifted much of its dominant logic away from the exchange of tangible goods (manufactured) things and toward the exchange of intangibles, specialized skills and knowledge, and processes (doing things for and with)', which they believe points marketing towards a logic that 'integrates goods with services and provides a richer foundation for the development of marketing thought and practice'. Based on Constantin and Lusch's (1994) distinction between 'operand resources' – resources on which an operation or act is performed to produce an effect – and 'operant resources' – resources which are employed to act on operand and other operant resources – Vargo and Lusch (2004: 3) argue that the service-centred dominant logic 'perceives operant resources as primary, because they are the producers of effects'. Besides, they propose knowledge – an operant resource – to be 'the foundation of competitive advantage and economic growth and the key source of wealth', a fact that is at the same time one of the foundational premises of the service-dominant logic (Vargo and Lusch, 2004: 9). As has been shown, knowledge – and specifically marketing knowledge – is also the key operant resource and the fundamental source of competitive advantage in knowledge-based approaches to marketing.

Vargo and Lusch (2004: 5–6) maintain that 'the service-centered view of marketing perceives marketing as a continuous learning process' and that 'a market-oriented and learning organization (Slater and Narver 1995) is compatible with, if not implied by, the service-centered model'. As the concepts of market-orientation and learning organization have greatly contributed to the evolution of knowledge-based concepts and indeed form essential building blocks for them, it is not difficult to see that the service-dominant logic and the knowledge-dominant logic of knowledge-based approaches to marketing are highly interrelated and mutually dependent. Indeed, as knowledge is the key operant resource and fundamental source of competitive advantage, a knowledge-centred view and efficient management of this resource are the *sine qua non* for effective marketing in the service-dominant logic.

This book has basically explored the role of knowledge and innovation in this new service-dominant logic by looking into the evolution of and the actual need for knowledge-based approaches to marketing for firms in the global knowledge economy. Indeed, I believe that in order to successfully adopt and make use of a service-dominant view, firms will have to adopt and apply knowledge-based approaches – a knowledge-dominant logic that supplements and supports the service-dominant logic so to speak. Taking heart from Gummesson's (2004b: 5) statement that '[t]he development of general marketing theory requires the integration of new lessons on a higher conceptual level than the existing theory, or to change the foundation of marketing theory', I propose that marketing theory needs to embrace this new knowledge-based logic for marketing.

The cases of HP CI, Schindler, Siemens, Toyota, Mazda, and Maekawa have helped to illustrate what knowledge-based approaches to marketing can look like in practice. In fact, facing the current global business environment and fierce competition, knowledge-based marketing has already become crucial as a determinant of corporate competitive advantage and as such a *sine qua non* for the six firms. Especially when introducing new products, entering new markets, or when trying to transform the whole business into a market- and customer-oriented organization, knowledge creation and transfer and intra- as well as inter-firm collaboration prove critical for the success of the projects. Therefore, applying knowledge management concepts and practices to the knowledge-intensive field of marketing and to marketing functions was particularly efficient and effective. Besides, the informant companies' emphasis on personalization strategies – supported by codification strategies using databases and intranet portals – and the knowledge of local staff showed that as large parts of marketing knowledge are tacit and hard to codify, face-to-face communication and the integration of local staff into marketing processes and decision-making are important factors for global marketing knowledge sharing that leads to successful marketing and sales achievements. In fact, as will be recalled, '[i]n a world where other firms are seeking to expand their market share, successful firms often can only stay ahead of the competition by exploiting new knowledge to offer improved products or processes that deliver new forms of added value to their customers' (Chaston, 2004: 155).

Finally, the cases and analysis have clearly underscored the essence of global knowledge creation and that it differs from and means much more than merely 'managing' knowledge (see also above). According to Takeuchi (2001: 315), knowledge management 'hit the West like lightning' at the beginning of the 1990s. We may well hope that the lightning (or enlightenment) of knowledge-based management strikes with at least the same impact since evolving to a knowledge-dominant logic for marketing will become more and more critical to corporate success and survival.

6.4 Conclusion

As Hansen and Nohria (2004: 22) correctly note, the ways for MNCs to compete successfully by exploiting scale and scope economies or by taking advantage of imperfections in the world's goods, labour, and capital markets are no longer as profitable as they once were, and as a result, 'the new economies of scope are based on the ability of business units, subsidiaries and functional departments within the company to collaborate successfully by sharing knowledge and jointly developing new products and services'. This statement strongly supports my call for knowledge-based marketing. At the same time, '[m]anagers and executives must strive towards meeting the slogan, "think globally and act locally" to be truly successful in managing knowledge across borders' (Desouza and Evaristo, 2003: 66), as in the era of globalization, 'a firm has to achieve global integration and local adaptation at the same time' (Nonaka and Toyama, 2002: 998).

The following four key conclusions can be drawn from this book:

(1) In an increasingly global business environment, the creation and transfer of marketing knowledge and intra-firm collaboration through knowledge-based approaches to marketing will become more and more crucial as a determinant for corporate competitive advantage and survival of firms.

(2) As marketing affairs are one of the most knowledge-intensive parts of a company, applying knowledge management concepts and practices to the field of marketing and to marketing functions will prove especially efficient and effective.

(3) As large parts of marketing knowledge are tacit and hard to codify, face-to-face communication and the integration of local staff into marketing processes and decision-making will be critical factors for global marketing knowledge sharing that leads to successful marketing and sales achievements.

(4) As no firm can be seen as isolated in the global network economy, relationship marketing and the co-creation of knowledge and value with other entities in the business ecosystem are increasingly important.

This book has presented a conceptual framework of knowledge-based marketing and highlighted essential processes of marketing knowledge co-creation with the main actors in the business ecosystem of global firms – customers, suppliers, competitors, business partners. As shown, traditional marketing approaches have focused overly on explicit knowledge and neglected the important role of tacit knowledge, specifically in international (cross-cultural) settings. This book's aim was to adjust this imbalance in the extant literature, and – drawing on real-life examples of knowledge-based firms – calls for a new knowledge-based marketing paradigm, with knowledge and knowledge co-creation being the key to sustainable competitive advantage in the global knowledge economy. We have seen that in the current global business environment, and faced with stiff competition, knowledge-based marketing is crucial for corporate competitive advantage – as demonstrated in the cases of Toyota, Mazda, Schindler, Siemens, HP, and Maekawa, for whom knowledge creation, transfer, and collaboration were essential for the success of their activities. Thus the application of knowledge management concepts and practices to marketing and its functions is particularly efficacious for firms such as these. Since much of marketing knowledge is tacit, to succeed in marketing and sales firms should encourage communication and collaboration and harness their employees' expertise in the processes of marketing and decision-making. Finally, all of the above mentioned companies can be seen as firms with a strong market orientation, possessing 'the basis for rapid adaptation to customers' manifest and latent needs, which may translate into superior new product success, profitability, market share, and, perhaps, sustainable competitive advantage' (Baker and Sinkula, 2005: 483). Even though it might be too early to identify and present real best practices, the 'next practices' (Prahalad and Ramaswamy, 2004a) quoted above show that some global leading companies are on their way to successfully implementing and leveraging knowledge-based marketing for competitive advantage.

Finally, as mentioned above, it is important to note that creating knowledge just for the sake of creating it will most likely not lead to sustainable competitive advantage. Knowledge-based management and the process of knowledge creation must aim at generating innovation and as a result higher value propositions to the firm's environment, specifically its customers but also all other stakeholders. Indeed, '[k]nowledge is a crucial enabler for innovation' (Ichijo, 2002: 478) and 'plays a critical role in innovation generation' (Hanvanich, Droge, and Calantone, 2003: 126; cf. also Cavusgil, Calantone, and Zhao, 2003), which is why Chaston (2004: 150) contends that innovation 'involves the application of knowledge to create new products and/or services'. It is therefore not surprising that market orientation and its link and relation with innovation has frequently been discussed in the relevant literature (Baker and Sinkula, 1999b; Calantone, Cavusgil, and Zhao, 2002; Darroch and McNaughton, 2003; Hurley and Hult, 1998; Vicari and Cillo, 2006) and that it is specifically customer knowledge which constitutes 'an important ingredient in innovation processes' (Gibbert, Leibold, and Probst, 2002: 467).

6.5 Managerial implications

This book has presented and discussed knowledge-based approaches to marketing management and used the cases of six outstanding global companies as vivid examples. But it is important to note that there is no silver bullet or single right approach. Indeed, depending on each company's individual circumstances, a particular knowledge-based approach to marketing will have to be developed and strategically managed. Nevertheless, based on the findings from the research project, I propose four important managerial implications:

(1) As knowledge has become a critical source for competitive advantage, marketing – and management in general – has to become knowledge-based.
(2) Marketing scholars and practitioners have focused too much on explicit marketing knowledge in the past. Combining and synthesizing both tacit and explicit knowledge and subsequently

leveraging holistic marketing knowledge is a *sine qua non* for corporate success and the source of innovation.

(3) Marketing knowledge needs to be co-created, not only inside the firm or across different units of a corporation but also together with other stakeholders, most importantly customers, but also with suppliers, partners, and competitors.

(4) Managers have to perceive their firms as interconnected in the global network economy, and thus have to take relationship marketing and the co-creation of knowledge and value with other entities in the business ecosystem seriously.

I strongly believe that systematically approaching and managing these issues and tasks will be a major challenge for corporations in the twenty-first century.

6.6 Limitations and implications for further research

Although carefully researched, documented, and analysed, the findings from the empirical study are subject to some limitations. First of all, the insights gained were derived and concluded from an exploratory study adopting a case study research design and are thus based on single – each probably rather unique – cases, even if this is exactly what case study research is all about (Stake, 2000). Indeed, the common limitations of generalizability of such field research are well documented (cf., for example, Eisenhardt, 1989; Hartley, 2004; Parkhe, 1993; Yin, 2003a), though analytic generalization – in contrast to statistical generalization – is possible (Hartley, 2004; Numagami, 1998; Yin, 2003a; cf. also Flyvbjerg, 2006a). Therefore, it would prove helpful to conduct further case studies of the companies presented in this book – longitudinal case studies (Leonard-Barton, 1990; Yin, 2003a) – and also of other firms in order to analyse knowledge-based approaches to marketing in different environments and under different conditions. I have used case studies and argued that they might not necessarily be seen as best practices but rather as next practices (Prahalad and Ramaswamy, 2004a). They were explanatory ones that used 'the force of example' but were probably not (yet?) what Flyvbjerg (2006a: 232) terms paradigmatic cases.

Furthermore, the marketing knowledge co-creation framework is based on and grounded in Nonaka's theory of organizational knowledge creation. But despite its development and advances over more than twenty years now, even Nonaka and Toyama (2003: 2) still conclude that 'it seems that we are still far from understanding the process in which an organization creates and utilizes knowledge' (cf. also Nonaka and Toyama, 2002) and the theory has also frequently been criticized in various aspects (for example, Glisby and Holden, 2003; Gourlay, 2006; Gueldenberg and Helting, 2007; Zhu, 2006). However, Gummesson (2002: xiii) is certainly right when he states that '[k]nowledge development in practical work in learning organizations as well as scholarly research can only humbly report progress and should not boast about conclusive results'. Besides, in viewing all knowledge as tentative, researchers have to train themselves to listen to reality without preconceived ideas. At a later phase the results can be compared with existing concepts and theory and will thus proceed as an interplay between the inductive and the deductive (Gummesson, 2005: 322–3).

Next, there are a few important issues I have not been able to discuss but which should definitely be included in future research: the role of power and value in knowledge-based management, the issues of motivation and incentives to create and share knowledge (for example, Gupta and Govindarajan, 2000b; Natter, Mild, Feuerstein, Dorffner, and Taudes, 2001; Nonaka, Toyama, and Nagata, 2000b: 12–13; Nonaka and Toyama, 2002: 1004, 1005; Osterloh and Frost, 2002; Osterloh, Frost, and Frey, 2002; Taudes, Trcka, and Lukanowicz, 2002), micropolitical issues (for example, DeMarco and Lister, 1999; Pfeffer, 1992), and research on social capital (Nahapiet and Ghoshal, 1998).

Finally, I take heart from Gummesson (2001: 44): 'To generate new knowledge in marketing, scholars should be guided by curiosity and the search for truth. Science must take risks and make mistakes; it must be entrepreneurial, not bureaucratic.' In writing this book, I was certainly guided by curiosity and the search for truth, a new truth for international marketing. I have probably also made mistakes along the way and this book cannot provide a final answer. But it is my sincere hope that it provides us with the opportunity to discuss, learn, and co-create a new dominant logic for doing business in the knowledge economy of the twenty-first century.

Afterword: Knowledge-based Marketing, So What?

Dr Charles M. Savage

How quickly we become comfortable with the things that worked yesterday. We in marketing have the language down pat, the routines are well established, and the questions memorized. Moreover, as marketers, we know what is needed. Unfortunately, engineering is not listening and sales are not really building upon our collateral.

So, why should we want to learn something about knowledge-based marketing? Our corporate headquarters is able to send messages to all parts of the world. And of course, the customers within our home market are the most sophisticated, so we really know what they and the others need, no matter where they are located.

When we do our strategic planning for the year, using both Porter's Five Forces and SWOT analysis, everything comes out as we'd like. Well, this was the case until one of our Asian marketing managers brought Kim and Mauborgne's *Blue Ocean Strategy* book to one of our meetings. He asked us, 'Are we really content to remain in a red ocean, bloody with competition?' 'Or,' he added, 'would we rather be in a blue ocean where the competition is irrelevant?' As we all agreed that the blue ocean is better, he smiled. 'If this is the case, we are going to have to take a whole new approach to marketing, both to our existing customers AND to those who are not yet our customers.'

Gradually, the comfort of our little marketing world began to feel uncomfortable. We turned to our friend and challenged him to explain more. Even though it is not a direct competitor, he used Maekawa Manufacturing Ltd. as an example. Instead of co-ordinating everything tightly from headquarters, they work with a decentralized structure which is project-based. Instead of telling their customers what they need, they listen. And this listening is not just to the words of the customers, but to what is behind the words. Moreover, they listen not only to the thoughts of the customers, but also to their feelings. They have learned that these tacit thoughts and feelings are the clue to a successful co-creation of new and appreciated market solutions. In short, they have learned to co-create and co-innovate WITH their customers.

One of the old-timers couldn't hold back any longer. 'If we listen to ALL our customers, we will get hopelessly lost. This is wrong. After all, they are paying us to TELL our customers what works best for them.'

Our Asian manager had anticipated this type of response, but instead of getting angry, he quietly smiled and said, 'You are absolutely right, there was a time when this approach worked, and it worked very well. I used to look forward to receiving your material and would use it without any changes with my customers in South East Asia. But I am sorry; this approach does not work any longer.'

'What do you mean?' replies the old-timer. 'We haven't been standing still. Aren't you impressed with the new CRM system we have put in, and our effort to do "one-to-one" marketing?'

Our Asian friend gently scratched his head. 'Well yes, it helps, but until we learn to actively learn not only with and between ourselves in the various regions, but with our customers, we will never be able to use the real talents of our engineers, our supply chain experts and our sales force. It is all about KNOWLEDGE, both tacit and explicit. And it is also about our ability to be open, honest, and REFLECTIVE together. Without asking the right powerful questions, without deep self-reflection and without listening ever so carefully to what is behind the words, we will miss many OPPORTUNITIES.'

The room was silent. The Asian manager could see the restlessness of the younger managers in the room. He asked them what they had learned from their colleagues in other parts of the world. Suddenly a tidal wave of frustrations turned to excited expression. Finally, they all said, 'It is possible to share lessons learned in conversations with customers in all parts of the world, to identify the innovations they expect, and connect with the aspirations of those who are not yet our customers.'

Florian Kohlbacher has brilliantly captured the important shifts under way in international marketing. What is even more exciting is his ability to show how marketing can benefit not only by the SECI (Socialization, Externalization, Combination, Internalization) approach so carefully worked out by Professors Nonaka and Takeuchi, but also their deep understanding of *ba*, the Japanese approach to quiet open reflection, reflection within an organization internationally, and reflection with suppliers and customers. Through these approaches, marketing in international corporations takes on a new importance. Instead of 'telling' those who are not 'listening', by using a knowledge-based marketing approach, marketing can now discover the deeper patterns upon which to build dynamic and expanding international businesses.

As someone who has been actively involved in knowledge management, intellectual capital, and knowledge economy developments, I deeply appreciate Florian's ability to bring together the best insights from many disciplines and illustrate them by in-depth case studies on HP, Schindler Elevators, Siemens, Toyota, Mazda, and Maekawa.

I can only predict that the wise international marketing department will not only study this book carefully, but they will ask their brightest marketers to develop a customized action work plan to guide them in better using their explicit AND tacit knowledge (and feelings). Those that move first will likely be able to reposition themselves in blue oceans of exciting opportunities, especially as they actively co-create their futures with their suppliers and customers.

Florian has opened an exciting door for all of us in international marketing; it is now up to us to walk through it. 'So what' will we do next? Perhaps by acknowledging Florian's valuable work, we can become our company's heroes. The choice is up to us!

President of Knowledge Era Enterprising International
Munich, Germany

Appendix: Research Methodology and Empirical Research

> Empirical research advances only when it is accompanied by theory and logical inquiry and not when treated as a mechanistic or data collection endeavor. (Yin, 2003a: xv)

This book is the result of an ongoing, continuous, and reiterative research process over two and a half years (April 2004–October 2006). Literature review, empirical data collection, theoretical reasoning, paper writing, and discussion with scholars and practitioners constantly alternated and overlapped in the process. This process actually resembles the SECI process (cf. Chapter 3.4.1) and its spiral and helped to continuously refine insights and knowledge gained.

As mentioned in Chapter 2.1, this book is essentially about knowledge and knowledge creation. But its aim is not only to illustrate, analyse and discuss knowledge-related processes in organizations but also to create new knowledge, that is, amend and extend existing theory and even build new theory. Gummesson (2004b: 3) in his treatise on the practical value of adequate marketing management theory, mentions that Vedic philosophy 'treats knowledge as a blend of three interacting elements: the process of knowing (methodology), the knower (the researcher) and the known (the results)' and that all three are needed in knowledge generation (cf. also Gummesson, 2001). In Chapters 4.1.2 and 4.2.1 I explored the nature of marketing knowledge and showed that the extant literature distinguishes between academic and practical marketing knowledge. In conducting empirical research and writing this book I am trying to blend the two and form them into holistic marketing knowledge, a marketing knowledge that has yet to be generated. One important role in generating this knowledge is the empirical research I conducted and the methodology used in the research. Indeed Probst (2002), for example, sees case-writing as a knowledge management – collecting, creating, transferring, and retaining knowledge – and organizational learning tool. This process of knowledge creation is, in a sense, also similar to the process of organizational knowledge creation discussed in the preceding chapters (specifically 3.4 and 4.2.3). Even though I am obviously not a research organization, but a single researcher by myself, the process of generating new marketing knowledge involved also other people and organizations, most notably my supervisors and other academic scholars, the people interviewed, the organizations I worked for or did participant observation in, and so on. Figure A1 shows the process of the research project and the dashed spiral illustrates the SECI process of marketing knowledge creation. Gummesson (2003a: 485) also speaks of a 'hermeneutic spiral', to stress the research as a dynamic process. The spiral also highlights the iterative character of the research, as you 'search again and again and again, just as the term says: re-search, re-search, re-search' (Gummesson, 2001: 29). Put differently: 'Through further theory generation in never-ending iterations we gain a spiraling effect and build a helix of continued development of knowledge' (Gummesson, 2001: 40).

This chapter deals with the methodological foundation of the book as well as the data collected and the methods used. I follow Gummesson's (2003a: 486–7) 'research edifice' (cf. Figure A2) and shall start from the basement, the foundations for research. Section A.1 introduces the general research paradigm underlying this book: explorative, qualitative research in marketing, which served as a framework and overall guideline of the whole research project. Indeed, '[a]ll research starts in the basement with the researcher's paradigm and preunderstanding' (Gummesson, 2003a: 486).

With section A.2, I will enter the middle floors of the research edifice: data generation and analysis. Chapter 5 is like an elevator, constantly moving between the middle floors and the penthouse, that is, the outcome. Finally, Chapter 6 is also situated at the penthouse, as it also discusses the results of the research project, their meaning, theoretical and managerial implications as well as limitations and recommendations for future research.

A.1 Qualitative research in marketing

> Let's stop fooling ourselves: All research is interpretive! (Gummesson, 2003a: 482)

There has been an ongoing debate on the appropriateness of different approaches and methods in social research. As a matter of fact, many authors point to the heated discussions, sometimes even 'wars' (the so-called 'paradigm war'), between the adherents of quantitative (so-called 'QUANs') and qualitative research (so-called 'QUALs') designs (for example, Bryman, 2004; Tashakkori and Teddlie, 1998). One main characteristic of this dispute seems to be the dichotomous way in which qualitative and quantitative research (methods) are presented as well as the resulting strict contraposition of the two (cf. also Bryman, 2004: 57–9). But it is also important to note that 'the sharp separation often seen in the liter-

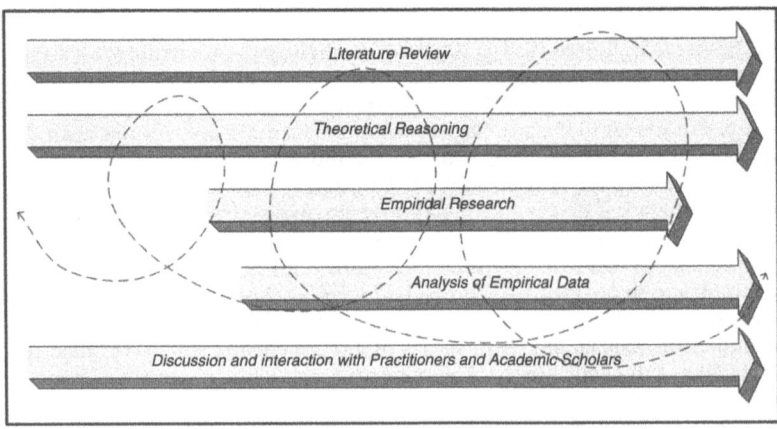

Figure A1 Process of research project and marketing knowledge creation (author's own illustration)

Figure A2 The research edifice (Gummesson, 2003a: 486)

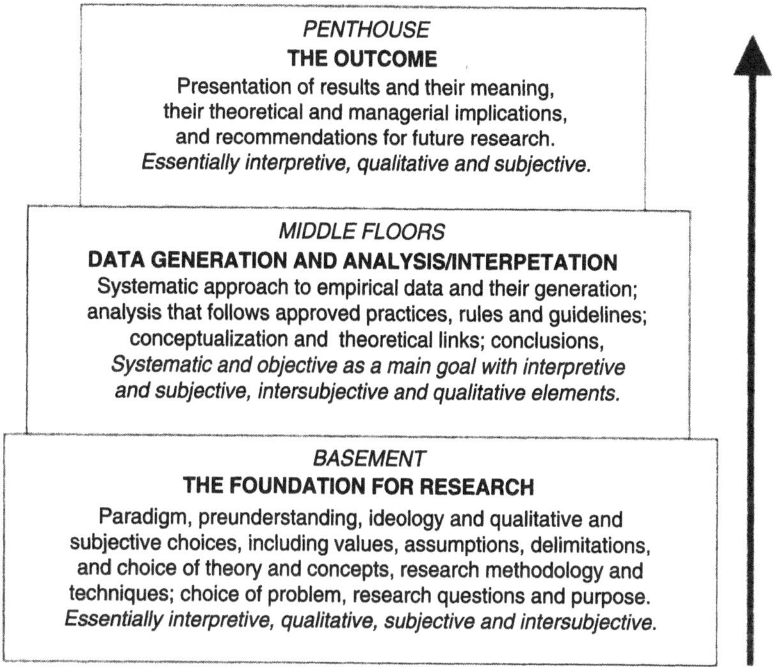

```
┌─────────────────────────────────────────────────────┐
│                     PENTHOUSE                          │
│                    THE OUTCOME                         │
│         Presentation of results and their meaning,     │
│        their theoretical and managerial implications,  │
│          and recommendations for future research.      │
│      Essentially interpretive, qualitative and subjective. │
└─────────────────────────────────────────────────────┘
┌─────────────────────────────────────────────────────┐
│                    MIDDLE FLOORS                       │
│       DATA GENERATION AND ANALYSIS/INTERPETATION       │
│      Systematic approach to empirical data and their generation; │
│     analysis that follows approved practices, rules and guidelines; │
│        conceptualization and  theoretical links; conclusions,  │
│       Systematic and objective as a main goal with interpretive │
│        and subjective, intersubjective and qualitative elements. │
└─────────────────────────────────────────────────────┘
┌─────────────────────────────────────────────────────┐
│                      BASEMENT                          │
│              THE FOUNDATION FOR RESEARCH               │
│        Paradigm, preunderstanding, ideology and qualitative and │
│      subjective choices, including values, assumptions, delimitations, │
│         and choice of theory and concepts, research methodology and │
│      techniques; choice of problem, research questions and purpose. │
│     Essentially interpretive, qualitative, subjective and intersubjective. │
└─────────────────────────────────────────────────────┘
```

Source: Copyright E. Gummesson (2003)

ature between qualitative and quantitative methods is a spurious one' (Flyvbjerg, 2006a: 241). Gummesson (2003a: 482) even speaks of a 'pseudo-conflict between quantitative and qualitative approaches'.

According to Spender (1996a: 72), 'the objective of positivist research is the development of a coherent abstract representation of the world out there' while the focus of interpretive research is 'on the ways in which we attach meaning to our experience'. Cassell and Symon (1994) contend that qualitative methods are more appropriate than quantitative methods to research questions focusing on organizational processes, as well as outcomes. One reason for this is that quantitative studies focus on the measurement and analysis of causal relationships between variables, not processes. Many scholars distinguish between explicit and tacit knowledge (see above and cf. also Spender's (1996b: 49–52) discussion of different types of knowledge) and Nonaka and Takeuchi's (1995) spiral of knowledge illustrates the process of creating knowledge in organizations through the interaction between tacit and explicit knowledge. Spender (1996a) emphasizes the contrast between research methods appropriate to explicit types of knowledge and those appropriate to implicit types, which according to him is also the contrast between the positivist and interpretive methods.

Furthermore, according to Spender and Grant (1996), the movement towards a knowledge-based theory of the firm was also influenced by an increasing criticism of the excessive abstraction and quantification in management education. The following quote exemplifies this well:

> The underlying paradigm of management research came under threat, reflecting an academy-wide shift away from the pursuit of a natural science type of organizational theorizing towards a richer and more complex framework that included rather than excluded people and their idiosyncrasies, culture and history – and their knowledge and skills. (Spender and Grant, 1996: 6)

In Chapter 3.3 we have seen that the knowledge-based view of the firm deals with the subjective elements of management, such as management vision, the firm's value system, and the commitment of employees even though many management scientists have avoided dealing with the subjectivity of humans. Since humans are both objects and subjects of research at the same time, research in social science cannot be free from subjective factors (Nonaka and Toyama, 2005).

Qualitative research in marketing – which traditionally has been dominated by quantitative approaches – has gained more and more attention in recent years (cf., for example, Buber, Gadner, and Richards, 2004; Carson, Gilmore, Perry, and Grønhaug, 2001; Gummesson, 2001, 2003a, 2004b, 2005, 2006). Indeed, research in marketing 'too often regresses to simplistic surveys without in-depth reflection on the mechanisms being studied' (Gummesson, 2004b: 4) and 'academic praise of the supremacy of quantitative measurements shuts out most of marketing reality' – specifically tacit knowledge – and 'hence the creation of more valid and general marketing theory' (Gummesson, 2001: 18). Indeed, in a statistical survey, for example, tacit knowledge remains just that – tacit' (Gummesson, 2001: 32). Besides, 'positivistic, theory testing research retards theory development in marketing' (Perry and Gummesson, 2004: 313).

For Gummesson (2005: 322) the 'overarching approach' is case study research, which is 'systemic and holistic, aimed to give full and rich accounts of the relationships and interactions between a host of events and factors'. Indeed, it 'takes a systemic, holistic stance recognizing reality as it is, not just settling for descriptions but adding value through conceptualization. It does not assume away complexity, chaos, ambiguity, fuzziness, uncertainty and dynamic forces for the convenience of the researcher and his or her analysis' (Perry and Gummesson, 2004: 315; cf. also Gummesson, 2003a).

A.2 Data and method

Section A.1 has briefly outlined the methodological foundation of this research project in a general manner. This section builds from this foundation and tries to explain the data collected and the methods used in more concrete terms.

A.2.1 Exploratory study – explanatory case studies

According to Schlegelmilch and Chini (2003: 220–1), even though 'marketing functions lend themselves particularly well for an investigation of knowledge transfer within MNCs', 'there is a dearth of research on knowledge transfer in the field of marketing'. This disconcerting gap in the fields of knowledge management and marketing has also become evident in conducting the literature review and was discussed in Chapter 4. It was exactly this gap, together with my own professional experiences in the field of marketing, that induced me to set out on a comprehensive empirical study of knowledge-based marketing in the first place. Besides, focusing on one particular type of knowledge – viz. marketing knowledge – helps to make the research become more efficient and effective because it is possible to concentrate on one part of the company and ask more concrete questions, rather than merely 'talking' about knowledge in general.

This state of affairs means that there is no accepted theory of marketing knowledge and knowledge-based marketing (cf. also Kohlbacher, Holden, Glisby, and Numic, 2007).[39] There being no theory to test, this particular contribution can do no more than attempt to generate theory from the explorative empirical study and the six case investigations that are introduced in this book. Indeed, 'before a theory can be validated, it must be constructed' (Patton and Appelbaum, 2003: 65). I recognize the limitations of building theory from case studies, but take heart from Eisenhardt's (1989: 536) encouragement that breakthroughs can be possible if one proceeds as if there is 'a clean theoretical slate'. As I hope to show, the particular cases represent a 'real-life method of inquiry ... [which] ... may be a nouveau solution' (Patton and Appelbaum, 2003: 69). This book is an attempt to break away from this rather tricky stranglehold and to attain 'a nouveau solution' for marketers who, like myself, are engaged in the quest of developing a knowledge-based concept of marketing. My approach focuses on knowledge co-creation with customers – as well as other stakeholders or entities in the business ecosystem of the firm – and the global transfer – or re-creation – of marketing knowledge.

Qualitative research is particularly useful for exploring implicit assumptions and examining new relationships, abstract concepts, and operational definitions (Bettis, 1991; Weick, 1996). According to Yin (2003a: 2) 'the distinctive need for case studies arises out of the desire to understand complex social phenomena' because 'the case study method allows investigators to retain the holistic and meaningful characteristics of real-life events', such as organizational and managerial processes (cf. also Kohlbacher, 2005). In fact, '[o]rganizations constitute an enormously complex arena for human behavior' (Dubin, 1982: 379) and case studies seem to be the preferred strategy when 'how' or 'why' questions are being posed when the investigator has little control over events, and when the focus is on a contemporary phenomenon within some real-life context (Yin, 2003a). In such a setting, case studies are explanatory ones, that is, they present data on cause–effect relationships, explain how events happened, and extend theoretical understandings (Yin, 2003a, 2003b). They make use of the 'power of example' (Flyvbjerg, 2001, 2006b).

The explanatory case studies are about critical projects or incidents in organizations, such as the exchange of critical customer, market, and product knowledge (cf. the HP CI case), the market introduction of a new product (cf. the Schindler case), the entry into a new market (cf. the Toyota case), the development of a new car – or new generation of a car – (cf. the Mazda case), the collaboration with a customer (cf. the Maekawa case), or the introduction of a new strategic approach (cf. the Siemens case). A critical case can be defined as 'having strategic importance in relation to the general problem' (Flyvbjerg, 2006a: 229). Critical incident technique (cf., for example, Chell, 2004) is often a suitable method for researching such critical cases or incidents. Moreover, it has been used to measure competencies and to identify tacit dimensions of knowledge acquired in solving real-world problems (Sternberg, Forsythe, Hedlund, Horvath, Wagner, Williams, Snook, and Grigorenko, 2000).

Hartley (2004: 323) states that case study research 'consists of a detailed investigation, often with data collected over a period of time, of phenomena, within their context', with the aim being 'to provide an analysis of the context and processes which illuminate the theoretical issues being studied'. Indeed, qualitative research approaches have been identified as offering 'holistic perspectives on phenomena which cannot be achieved otherwise' (Sinkovics, Penz, and Ghauri, 2005: 32), which is also why case studies have an important function in generating hypotheses and building theory (cf., for example, Eisenhardt, 1989; Hartley, 2004; Quattrone, 2006). Indeed, according to Patton and Appelbaum (2003: 67), the 'ultimate goal of the case study is to uncover patterns, determine meanings, construct conclusions and build theory'. Last but not least, case studies have been recognized as important and useful in knowledge creation (Probst, 2002; Remenyi, Money, Price, and Bannister, 2002) and for both theoretical and practical knowledge management (Probst, 2002) and marketing (Carson, Gilmore, Perry, and Grønhaug, 2001; Gummesson, 2003a, 2005; Perry, 2004).

This book presents theorizing that emerges from and is intertwined with empirical research. Similarly to the empirical research described by Yanow (2004: S11), the research unfolded and advanced in a process that tacks back and forth between empirical data and theorizing (cf. also Figure A1). The explorative or exploratory study which involved qualitative expert interviews served mainly as a means to identify and purposefully select critical cases for in-depth case studies (cf. A.2.2 and Chapter 5) but also to identify important research issues, to generate hypotheses or research propositions. These in-depth case studies were then conducted as explanatory cases to illustrate and sustain the argument for knowledge-based approaches to marketing. This actually makes the case material central to the theorizing process, that is, it was generative for the theorizing – even more than the explorative study. But in the writing up (Chapter 5), the cases serve more as an illustration of the theoretical arguments. Last, but not least, it is important to note that there was also continuous tacking back and forth between exploratory study and explanatory case study research. Indeed, this research project involved two important steps of empirical research, namely the explorative study and the explanatory case studies. But these two steps were not conducted in linear sequence. Obviously, the research project started with the explorative study and its qualitative interviews in order to learn about and identify knowledge-based approaches to marketing. Based on these insights, a number of companies were selected for in-depth case studies in order to analyse and exemplify particularly critical cases. At the same time, I kept conducting the explorative part of the study and kept interviewing managers, employees, and other experts as part of the purposeful sampling process (chain sampling).

Gummesson (2005: 318–19) sees conceptualization and contextualization as two key issues in theory generation and deems them to be interwoven and stressing different aspects of theory generation. He contends that 'concepts are needed, and in times of major changes new concepts – reconceptualisation – are urgently needed' and that contextualization refers to 'the need to place single data in a broader context, that is, generate theory' (Gummesson, 2005: 318). Moreover, theory orders data in a context and a theory is a roadmap and a good roadmap makes it possible to navigate in a territory that is unknown to the traveller (ibid.). But new theory can also develop from new interpretations and innovative combinations of extant theory. Finally, theory generation, 'moving from raw data and description to conceptualisation and contextualisation, may be the most valuable contribution a scholar can offer', even though as researchers in marketing 'we are rarely if ever innovators; we rather start out as observers and messengers' (Gummesson, 2005: 319). However, it is not enough to be reporters of events. We have to add value to the phenomena we present; that is what scholarship is all about (ibid.). Therefore, the outcome of successful theory construction is 'to conceptionalize a field, generalize beyond the mere description of events, and make it more intelligible and manageable' (Gummesson, 2002: 31).

Last but not least, it is also important to keep in mind Flyvbjerg's (2006a: 223) caveat that 'there does not and probably cannot exist predictive theory in social science' and that social science 'has not succeeded in producing general, context-independent theory and, thus, has in the final instance nothing else to offer than concrete, context-dependent knowledge'. As a matter of fact, case studies are 'especially well suited to produce this knowledge' (ibid.).

A.2.2 Sampling

As for sampling, I opted for purposive sampling (purposeful sampling) and theoretical sampling (for example, Glaser and Strauss, 1967; Gummesson, 2001). The former is essentially strategic and entails an attempt to establish a good correspondence between research questions and sampling, as the researcher samples on the basis of wanting to interview people who are relevant to the research questions (Bryman, 2004; Patton, 2002). In case study research, for example, the sample is 'theoretical and purposeful – find the cases that give a maximum of information – and guided by saturation – stop when the new information of additional cases approaches zero' (Gummesson, 2003a: 488). According to Patton (2002: 230, original emphasis), the 'logic and power of purposeful sampling lie in selecting *information-rich cases* for study in depth', with information-rich cases being 'those from which one can learn a great deal about issues of central importance to the purpose of the inquiry'. In fact, '[s]tudying information-rich cases yields insights and in-depth understanding rather than empirical generalizations' (Patton, 2002: 230). This kind of sampling can also be called 'strategic sampling' and the strategic choice of case may greatly add to the generalizability of a case study (Flyvbjerg, 2006a: 226). Theoretical sampling entails sampling interviews until your categories achieve theoretical saturation and selecting further interviewees on the basis of one's emerging theoretical focus (cf., for example, Bryman, 2004; Glaser and Strauss, 1967; Guest, Bunce, and Johnson, 2006; Patton, 2002; Strauss and Corbin, 1990). Last, but not least, it is important to note that locating critical cases 'requires experience, and no universal methodological principles exist by which one can with certainty identify a critical case' (Flyvbjerg, 2006a: 231).

The companies studied for this book were selected purposefully by choosing firms and cases that seemed to be most appropriate to provide insights into knowledge-based management processes and specifically knowledge-based approaches to marketing. These companies and cases were identified through a review of the relevant literature and widely recognized knowledge management studies such as the MAKE (Most Admired Knowledge Enterprise) award[40] (cf., for example, English and Baker, 2006) Furthermore, a kind of 'snowball' or 'chain sampling' approach (Patton, 2002: 237) also proved helpful with knowledge management experts in corporations as well as research institutes and universities indicating other key informants or critical cases to me. This approach fits with Easterby-Smith and Araujo's (1999) call for studies which develop theory from practice and which use a small sample of in-depth cases, which focus on micro-practices within organizational settings and which study processes and competencies leading to learning outcomes. Of the thirty-five companies, nine were purposefully selected to conduct in-depth case studies. Of these nine, six were purposefully selected to serve as explanatory cases studies of knowledge-based marketing as proposed in Chapter 4.2. These case studies are reproduced in Chapter 5.

Finally, the interviewees and key informants were also sampled purposefully. The main target were top managers, and middle managers and employees in charge of marketing, product management, and knowledge management. In total, qualitative interviews with 116 top executives, middle managers, and selected employees in thirty-five different companies were conducted in 2005 and 2006, mainly in Japan (90) but also including supplementary interviews – where appropriate and necessary – in selected countries, namely Austria (9), Czech Republic (5), Germany (9), Switzerland (1), and China (2). Two key informants (see below) were interviewed twice, resulting in data from a total of 118 interviews. The interviews can be separated into interviews for the explorative study, interviews for the nine case studies, and auxiliary interviews for the case studies. In the latter case, the interviews were either not directly related to the case researched or the interviewees were working for a different organization, but they had some relation with or special insights into the project/case researched and hence added value to the case study. Figure A3 gives an overview of the number of interviews conducted. More detailed information will be provided in section A.2.3.

A.2.3 Data generation and method

Following Gummesson (2003a: 486), I use the term 'data generation' rather than 'data collection', as 'data in social settings are not objects that are ready for collec-

	Explorative study	Nine case studies (main)	Nine case studies (auxiliary)	TOTAL
Number of interviews	35	53	30	118

Figure A3 Number of interviews conducted as relevant for sampling

tion', but instead 'data are generated, meaning that they are the creation of the researcher in interaction with, for example, a respondent in an interview'.

A.2.3.1 General description

The research methodology involved triangulation among a variety of different sources of data (cf., for example, Bryman, 2004; Parkhe, 1993; Wolfram Cox and Hassard, 2005) including the conducting of both formal and informal on- and off-site interviews with managers as well as scholars and other experts in the field, analysis of archival materials such as company internal documents as well as articles in the business media, and an evaluation of existing case studies and other relevant literature (Yin, 2003a). In sum, primary sources, secondary sources, and tertiary sources were used and triangulated.

The explorative study, which was the starting point of the empirical part of the book project, was carried out in the form of qualitative interviews or expert interviews. This means that open questions are used and the interviewees are encouraged to talk as freely and undisturbed as possible, with the interviewer not trying to structure the course of the interview. This is in order to follow the interpretive approach to social knowledge, which recognizes that 'meaning emerges through interaction and is not standardized from place to place or person to person' (Rubin and Rubin, 1995: 31). Therefore, in the course of the qualitative interviews, semi-structured questions in accordance with the theory of organizational knowledge creation and enabling were employed, but the interview partners could nevertheless answer openly and lead the interview mostly. Interview guidelines were used and each interview guideline was prepared specifically before each single interview. In addition to fundamental key topics,[41] which always remained in place, company- or interviewee-specific questions or topics were prepared, depending on the background information available beforehand. Unless no permission was given or unless it seemed inappropriate for other reasons, all interviews were recorded and authentically transcribed. After analysis and development of the case studies, the drafts were examined by key persons from the respective companies and their feedback was used for further refinement. Indeed, working closely together with the informant companies has led to a knowledge co-creation process through the co-creation of the case studies – especially HP and Siemens. As mentioned above, in total, qualitative interviews with 118 top executives, middle managers, and selected employees in thirty-five different companies have been conducted.

In addition to the interviews participant-observation (Bryman, 2004; Gillham, 2000; Patton, 2002; Waddington, 2004; Yin, 2003a) was used for the case studies of Schindler and Siemens (cf. below). This is consistent with Gummesson's (2001, 2005) concept of 'marketing anthropology' in interactive research. According to Flyvbjerg (2006a: 236), research can be seen as a form of learning and 'the most advanced form of understanding is achieved when researchers place themselves within the context being studied' because it is only in this way that researchers can understand the viewpoints and the behaviour which characterize social actors. In fact, as Osland and Cavusgil (1998: 200–1) have noted, '[i]n depth field research methods enable researchers to gain a rich understanding of respondents' perspectives, often providing insights that the researcher would not have uncovered from structured questionnaires used in traditional surveys'.

Using participant-observation and actually frequently going beyond it, basically led to action research. As a matter of fact, action research in marketing has received considerable attention and discussion in recent years (cf., for example, Ballantyne, 2004; Carso, Gilmore, Perry, and Grønhaug, 2001; Gummesson, 2001, 2005; Kates and Robertson, 2004; Perry and Gummesson, 2004). According to Patton (2002: 221), action research 'aims at solving specific problems within a program, organization, or community' (cf. also Bryman, 2004; Greenwood and Levin, 1998; Heller, 2004; Lewin, 1946; Perry and Gummesson, 2004), and Gummesson (2005: 323) argues that action research 'entails dialogue and reflection based on data from experience through active involvement in the process being studied'. He further elaborates that:

> Management action research is an application to the study of business phenomena, and a subdiscipline could be named marketing action research. The action researcher does scholarly research and is both an academic researcher and either a marketing practitioner or an external consultant. His or her purpose is twofold: to contribute to science and to help solve a practical problem. By being involved, the object of study creeps under the skin of the researcher in a way that is not possible in the study of documents or in interviews, even in participant observation. The access is as close as can be, and tacit and embedded knowledge can be uncovered. (Gummesson, 2005: 324).

A.2.3.2 *Overview of the informant companies and number of interviews*

Figure A4 gives an overview of the informant companies and the number of interviews conducted for the case studies. As mentioned above (A.2.2), of the nine case studies only six were purposefully selected for reproduction and analysis in this book. Given that the final sample contains two automotive companies as well as the complex supplier relationships and harsh competition in the automotive industries, twelve auxiliary interviews were conducted with other suppliers and OEMs to gain a more complete picture. This is especially crucial since case 4 is about knowledge co-creation with competitors and suppliers. The firms involved were: Bosch Japan (4, additional 2 interviews were specifically for TPCA), BWM Japan (1), Denso (4), Mitsubishi-Fuso/ Daimler Chrysler (2), and Nissan (1). The following sections explain the interviews conducted at each informant company for the case studies.

A.2.3.3 *Hewlett-Packard Consulting & Integration (HP CI)*

For this case study, a total number of five qualitative interviews were conducted in 2005 and 2006: two with the lead knowledge adviser and a knowledge manager at the knowledge management department of HP CI in Vienna, and two with the head of the knowledge management department of HP CI Japan and his subordinate staff. One additional interview was conducted with a Senior Consultant at HP CI Japan. The two interviews in Vienna were conducted in German, the three interviews in Japan in Japanese. As for the head of the knowledge management department of HP CI Japan – who is by now a professor at Hamamatsu University, Department of Management Information Systems – we have met on numerous further occasions to discuss knowledge management issues, attend knowledge management-related events, jointly con-

No.	Informant company	Interviews for case study	Auxilliary interviews for case study	Interview location(s)
1	HP CI Japan	5	–	Japan, Austria
2	Schindler Elevator	9	–	Japan, Hong Kong/Shanghai, Austria
3	Siemens	8	11 (2 for TPCA)	Japan, Germany, Austria, Switzerland, China
4	Toyota Peugeot Citroën Automobile (TPCA)	5	7	Czech Republic, Japan
5	Mazda Motor Corporation	5	0	Japan
6	Maekawa Manufacturing	5	0	Japan
	TOTAL	**37**	**18**	**55**

Figure A4 Informant companies and number of interviews for the case studies

duct interviews, and so on, and maintain frequent contact as well. This collaboration finally also led to the co-creation of a joint article on HP CI Japan (Kohlbacher and Mukai, 2007), on which the case study is based.

A.2.3.4 Schindler Elevator

The Schindler case is based on nine qualitative interviews with key persons in marketing and product management at the Competence Centre Escalator in Vienna and Schindler's subsidiary in Tokyo (including the President), as well as with the product line manager in charge of the market introduction project (see below) at the Asia-Pacific headquarters in Hong Kong. The interviews were conducted in 2005 and 2006 in German and Japanese, depending on the native language of the interviewee. I also used participant observation in addition to the interviews and worked on the research project as a part-time employee for five months at the Escalator Division of Schindler Elevator K.K., Japan from August to December 2005. This can also be seen as action research since I was directly involved in the new market introduction project described in the case study (5.3). In total, I worked intermittently for Schindler Elevator K.K. and Schindler Lifts and Escalators Ltd., Vienna in different marketing-related functions from 2002 to 2005, a fact that is important given Gummesson's (2001: 28) pronouncement to the effect that he sees the researcher 'as the number one research instrument' and that he feels 'at liberty to use [himself] and [his] experience as evidence'.

A.2.3.5 Siemens

At Siemens, a total number of twenty-one interviews were conducted. Of these, eight interviews were conducted with people directly involved in or in charge of the researched Siemens One case: four interviews at the Corporate Development Siemens One headquarters in Munich, Germany (in English), one with the Siemens One manager at Siemens Japan (in Japanese), two with the Siemens One manager and one key account manager at Siemens Austria (in German), and one with the Vice President and Siemens One manager at Siemens China (in German). In addition to that, eleven auxiliary interviews for the case study were conducted with people who had relevant information for the case or people involved in knowledge management activities at different divisions: four in Japan (in German), two in Austria (in German), four in Germany (in German) and one in Switzerland (in German). As mentioned below (A.2.3.6), two further interviews were conducted with people from the Competence Centre Automotive who were involved in the TPCA project. Finally, as in the Schindler case, an action research approach was taken by working on the researched projects as a part-time employee at the Siemens One department of Siemens K.K., Japan from January 2006 to September 2006 in order to conduct participant observation.

A.2.3.6 Toyota Motor Corporation/Toyota Peugeot Citroën Automobile Czech (TPCA)

According to Parkhe (1993: 228) 'inductive/theory-generating/idiographic research may provide a powerful stimulus that is particularly well suited for the current stage of evolution of IJV research'. Even though I acknowledge that IJV research has significantly advanced since this statement was made in 1993, I believe that a case study research strategy can help to provide the necessary stimulus and shed light on crucial issues such as learning and knowledge creation and enabling in IJVs, which are still not fully understood, especially in the peculiar case of two foreign competitors in a transitional economy.

I also followed Osland and Cavusgil's (1998: 200) recommendation to collect multiple-party perspectives, which is 'especially critical when examining international joint ventures that involve parent companies from dissimilar cultures'. Therefore interviews were conducted in 2006 with the first TPCA President who had already returned from TPCA to Toyota headquarters in Japan, as well as with key persons of both Toyota and PSA at TPCA in Kolín, including the President and Executive Vice President. In total five managers were interviewed, one in Japanese and the others in English. Moreover, additional insights and views were gained from interviewing key persons at TPCA suppliers and external Toyota experts in Japan, Germany, and the Czech Republic. Specifically interviews with two Toyota key account managers at Bosch Japan (parts supplier) and two key account managers and project leaders at Siemens Japan (supplier of factory automation equipment) and interviews with three independent Toyota experts – all working as consultants on their own account – have been conducted. This led to a total of five TPCA and seven auxiliary interviews. This research is the fruit of a collaboration with Kaz Ichijo and has been documented by a series of conference papers/articles in 2006 (for example, Ichijo and Kohlbacher, 2006a, 2006b, 2007; Rädler, 2006), which served as a helpful basis for the case study.

A.2.3.7 Mazda Motor Corporation

The Mazda MX-5 roadster case study research is the fruit of collaboration with Jiro Nonaka and involved qualitative interviews with the two project leaders of the product development teams of the three generations of MX-5 roadsters, Toshihiko Hirai and Takao Kijima in 2005. A second interview with Takao Kijima was conducted in 2006, as well as with the manager of the newly established knowledge management department. Additionally, an interview with Kentaro Nobeoka of Kobe University, Research Institute for Economics & Business Administration – who has seven years of experience as a product planner with Mazda Motor Corporation, where he was involved in project management teams for several different models, including the researched case – was conducted in 2006. All interviews were conducted in Japanese. Nonaka and Katsumi's (2006) Japanese article on the development of the roadster served as a very helpful base for researching and writing up this case study, and a preliminary version of the case study was presented at a research workshop (Nonaka, Kohlbacher, and Holden, 2006), where useful feedback was obtained.

A.2.3.8 Maekawa Manufacturing

The Maekawa case study involved a total of five qualitative interviews in 2006. In addition to two interviews with the President of the Maekawa General Research Institute, the former President and honorary Chairman, the head of the Corporate Communications Department and the general manager of the International Project Department were interviewed. Previous research on Maekawa, its management philosophy, and its particular project-based system has been published both in Japanese and in English. The former President and honorary Chairman of the company, Masao Maekawa, co-authored a book with renowned system theorist Shimizu Hiroshi of Tokyo University, called *From Competition to Co-creation* in 1998 (Shimizu and Maekawa, 1998) and more recently he published a book on his management and manufacturing philosophy (Maekawa, 2004). Research on Maekawa Manufacturing has been published in Japanese (for example, Tsuyuki, 2001b, 2006) and English (for example, Nonaka and Konno, 1998; Nonaka, Konno, and Toyama, 2001; Peltokorpi and Tsuyuki, 2006; von Krogh, Ichijo, and Nonaka, 2000; von Krogh, Nonaka, and Ichijo, 1997).

A.2.4 Quality criteria

For case studies, theory development as part of the design phase is essential, whether the ensuing case study's purpose is to develop or test theory, with theory development taking place prior to the collection of any case study data being an essential step in doing case studies (Yin, 2003a: 28–9). But depending on the depth and range of the extant literature, the initial focus of the case study may be quite focused or broad and open-ended. Therefore, and because the case study strategy is ideally suited to exploration of issues in depth and following leads into new areas of new constructions of theory, the theoretical framework at the beginning may not be the same one that survives to the end (Hartley, 2004: 328). Besides, theory development not only facilitates the data collection phase of the ensuing case study, but the appropriately developed theory also is the level at which the generalization of the case study results will occur. This role of theory has been characterized by Yin (2003a: 31–2) as 'analytic generalization' and has been contrasted with another way of generalizing results, known as 'statistical generalization' (cf. also Hartley, 2004; Numagami, 1998). As a matter of fact, a common concern about case studies put forward by their critics is that they provide little basis for scientific generalization (Flyvbjerg, 2006a; Yin, 2003a). Yin (2003a: 10) answers this as follows:

> case studies ... are generalizable to theoretical propositions and not to populations or universes. In this sense, the case study ... does not represent a 'sample,' and in doing a case study, your goal will be to generalize theories (analytical generalization) and not to enumerate frequencies (statistical generalization).

According to Flyvbjerg (2006a: 221), to understand why the conventional view of case study research is problematic, we need to grasp the role of cases and theory in human learning. The case study produces the type of context-dependent knowledge that research on learning shows to be necessary to allow people to develop from rule-based beginners to virtuoso experts, and in the study of human affairs, there appears to exist only context-dependent knowledge, which, thus, presently rules out the possibility of epistemic theoretical construction. Indeed, context-dependent knowledge and experience are at the very heart of expert activity and such knowledge and expertise also lie at the centre of the case study as a research and teaching method or, to put it more generally still, as a method of learning (Flyvbjerg, 2001, 2006a).

Furthermore, a major issue in designing case study research is the maximization of conditions related to design quality, that is, the criteria for judging the quality of research designs. They are basically the following four: internal validity, construct validity, reliability, and replicability and external validity (cf., for example, Eisenhardt, 1989; Leonard-Barton, 1990; Numagami, 1998; Yin, 1981, 2003a).

According to Gummesson (2005: 322), quality criteria for quantitative studies, such as reliability and representativeness, cannot in general be applied to case study research. As shown above, the sample is theoretical and purposeful, looking for cases that give maximum information, and preferably is guided by saturation, the point where no or little new information is added (Gummesson, 2005: 322). Therefore, by choosing information-rich cases and using the 'force of example' (Flyvbjerg, 2006a), representativeness – in a qualitative sense – should be obtained.

As for reliability, Yin (2003a: 34, 37) states that the objective is to demonstrate that the operations of a study – such as the data collection procedures – can be repeated, with the same results and that the goal is to minimize the errors and biases of a study. By recording the qualitative interviews (where permission was given) and using case study protocols, I tried to ensure a high level of reliability.

Construct validity is about establishing correct operational measures for the concepts being studied, and this test is especially problematic in case study research (Yin, 2003a: 35). Yin (2003a: 34, 36) offers three tactics to increase construct validity when doing case studies: the use of multiple sources of evidence, establishing a chain of evidence, and a review of the draft case study report by key informants. In order to ensure construct validity, I focused on the use of multiple sources (cf. above, A.2.1) and a review of the draft case study report by key informants. The former is part of triangulation and the latter is also called 'respondent validation' and both help to establish trustworthiness and credibility of the research (Bryman, 2004). Where possible and appropriate, key informants commented on the vignettes and depending on this feedback they were used for further refinement of the cases studies and/or analyses of them. The inclusion of both 'insiders' – the key informants – and 'outsiders' – myself and the readers – as well as the inclusion of archival company data and participant observation allowed for extensive data, researcher, and method triangulation, adding richness to the evaluation and interpretation of the cases, thereby enhancing the internal and construct validity of the conclusions drawn (Stake, 1995; Yin, 2003a). As mentioned above, this has also led to knowledge co-creation between the informant companies – that is, certain interviewees – and me.

According to Bryman (2004: 273) credibility also parallels internal validity. Internal validity is about establishing a causal relationship, whereby certain conditions are shown to lead to other conditions, as distinguished from spurious relationships (Yin, 2003a: 34, 36). In my case studies, causal relationships were not the main focus of research and as a result, internal validity was not the most important criterion. Last but not least, external validity is about establishing the domain to which a study's findings can be generalized (Yin, 2003a: 34, 37). It is important to note that while (quantitative) survey research relies on statistical generalization, case studies rely on analytical generalization (cf., for example, Eisenhardt, 1989; Hartley, 2004; Perry and Gummesson, 2004; Yin, 2003a). In analytical generalization, 'the investigator is striving to generalize a particular set of results to some broader theory' (Yin, 2003a: 37). Remaining concerns for external validity (particularly statistical generalizability) were traded off against the opportunity to gain in-depth insights, but multiple case analyses were used to ensure at least analytical generalizability due to the different industries studied (Eisenhardt, 1989; Yin, 2003a).

Numagami (1998) questions the possibility of establishing invariant laws for social phenomena and presents an argument for the case study method. He argues that the conditions under which an invariant law can be discovered are so stringent that the search of an invariant law should not be the main objective of management studies, and he contends that reliability/replicability and external validity are irrelevant not only for the case study, but for any method of management studies. Indeed, while the concepts and techniques for meeting the criteria for internal validity and construct validity have developed effectively, those for meeting reliability/replicability and external validity have not (Numagami, 1998: 2). But the latter two criteria 'are relevant only if the social researcher is searching for an invariant and universal law' (Numagami, 1998: 3). I am neither looking for any invariant law for knowledge-based approaches to marketing, nor do I believe there is one. Therefore, the two criteria of reliability/replicability and external validity are not really relevant for my research project.

Finally, as mentioned above, the case study has its own rigour, different but no less strict than the rigour of quantitative methods, and the advantage of the case study is that it can 'close in' on real-life situations and test views directly in relation to phenomena as they unfold in practice (Flyvbjerg, 2006a: 235). In this context, it is also important to note that proof 'is hard to come by in social science because of the absence of "hard" theory, whereas learning is certainly possible' and that in essence, 'we have only specific cases and context-dependent knowledge' (Flyvbjerg, 2006a: 224). Therefore, I see the results from this research project as a learning process and learning insights rather than hard proof of a theory or ideas.

Notes

1. Of course, the term 'knowledge-based management' is not my invention and eminent scholars such as Ikujiro Nonaka also tend to use this term instead of 'knowledge management'. Cf. also Chapter 3.3 on the knowledge-based theory of the firm.
2. As will be relevant in later chapters (e.g. 4.2.2.2) Gummesson (2003b) has given the dual labels of 'value society' (with focus on output) and 'network society' (with focus on input) to the contemporary economy.
3. Teece, Pisano, and Shuen (1997: 516) define dynamic capabilities 'as the firm's ability to integrate, build, and reconfigure internal and external competences to address rapidly changing environments'.
4. For a detailed critical discussion of knowledge-based approaches to the theory of the firm see Conner (1991, 1996); Eisenhardt and Santos (2002); Foss (1996a, 1996b); Kogut and Zander (1992, 1996).
5. I acknowledge the fact that the dichotomous distinction between Western and Japanese approaches to knowledge creation according to a focus on explicit or tacit knowledge might be a somewhat oversimplified generalization. I know of exceptions to this rule on both sides and believe that it is frequently the corporate rather than the national culture that plays the decisive role in this context. Nevertheless, Nonaka's theory is widely acclaimed and accepted – for critical literature see 6.6 – with the state-of-the-art literature in the field of knowledge management recognizing and agreeing on this distinction. Moreover, both my own empirical research experience, as well as the differences in the focus of the knowledge management literature in the West and in Japan, seem to suggest that there is indeed a tendency of Western firms to focus on explicit knowledge and for Japanese firms to focus on tacit knowledge.
6. Interestingly, the Japanese translation of Wenger and Snyder's (2000) *Harvard Business Review* article – published in the *Diamond Harvard Business Review* (August 2001), pp. 120–9 – has the title 'The Innovation Power of Ba'. The translator mentions in a short note that CoPs are the same as the concept of 'ba' and uses the term 'ba' as a translation of CoP throughout the article.
7. Even though the concept of absorptive capacity has been widely discussed and used in the extant literature (e.g. Cohen and Levinthal, 1990; Van den Bosch, Van Wijk, and Volberda, 2003; Zahra and George, 2002), it was not researched in the empirical research project underlying this book and will therefore not be discussed in greater detail here.
8. For the concept of 'co-creation' see, for example, Savage (1996).
9. Formerly known as Dorothy Leonard-Barton.
10. As a matter of fact, both are very well acquainted with each other. Ikujiro Nonaka has also pointed out some of the similarities to me on several occasions (personal communications with Ikujiro Nonaka, May–September 2006).

11. Though the main focus in this book will be Nonaka's work.
12. Compare Bell, Whitwell, and Lukas' (2002) approach to reviewing the literature on organizational learning and marketing, which identifies four different schools of thought in organizational learning, but fails to discuss marketing knowledge and knowledge management issues.
13. According to Baker and Sinkula (2005: 484), more than 100 studies since 1990 have looked at the market orientation–performance relationship.
14. Kumar, Scheer, and Kotler (2000) distinguish four orientations to marketplace: sales driven, market driven, customer driven, and market driving. They finally make a strong claim for market driving companies. Jaworski, Kohli, and Sahay (2000), on the other hand, see both the market-driven approach and the driving-markets approach as approaches to being market-oriented. I follow this latter notion.
15. The actual starting point was probably Day and Wensley (1988).
16. Drucker's dictum is from Drucker (1954).
17. Slater and Narver (1995) see market orientation as one of five critical components of the learning organization, with the others being entrepreneurship, facilitative leadership, organic structure, and decentralized strategic planning. A detailed discussion of these other components would go beyond the scope of this book, however.
18. For a detailed analysis, discussion, and definition of organizational memory see Walsh and Ungson (1991).
19. The market sensing capability is complemented by the capability of 'peripheral vision', which is much more than sensing and is also 'knowing where to look more carefully, knowing how to interpret the weak signals, and knowing how to act when the signals are still ambiguous' (Day and Schoemaker, 2006: 2; cf. also the special issue of *Long Range Planning*, 2004).
20. Note that this article appeared as early as in 1977.
21. Sinkula (1994) proposes a hierachy of market knowledge that encompasses seven levels of knowledge, but a detailed discussion would go beyond the scope of this book and be unnecessarily complicated.
22. As a matter of interest, Brodie, Uncles, Wierenga, Midgley, and Rossiter are all close friends and members of the Marketing Knowledge Project's international committee (Rossiter, 2002).
23. Hackley (1999: 722) further notes that this is a 'problematic feature not simply of marketing but of every practical discipline' (cf. also Polanyi, 1962, 1966).
24. For a discussion of non-*keiretsu* business networks of small- and medium-sized firms in Japan see Ibata-Arens (2004).
25. The Zaltman Metaphor Elicitation Technique, employed by Olson Zaltman Associates and its licensees, is a patented research method, US Patent Number 5,436,830.
26. Slater and Narver (1998; 1999), in a discussion of market orientation in a series of articles in the *Strategic Management Journal*, distinguish between two forms of 'customer orientation' that are frequently confused. The first, a customer-led philosophy, is primarily concerned with satisfying customers' expressed needs, and is typically short term in focus and reactive in nature. The second, a market-oriented philosophy, goes beyond satisfying expressed needs to understanding and satisfying customers' latent needs and, thus, is longer term in focus and proactive in nature.

27. Except for Seven Eleven Japan, the only two other cases researched and discussed by Nonaka and associates of firms that consciously take knowledge-based approaches to marketing seem to be Ryohin and Maekawa (Senoo, Akutsu, and Nonaka, 2001). Besides, in their study of branding capabilities, Akutsu and Nonaka (2004) use the theory of organizational knowledge creation and an extended notion of brand knowledge and redefine the brand-building method as the brand knowledge-creation process, with Sony serving as an illustrative example. Katahira, Furukawa, and Abe (2003), in their book *Beyond Customerism*, offer a range of Japanese companies that go beyond mere customer focus and can, in their own particular way, be seen as taking knowledge-based approaches to marketing.

28. In contrast to that, cf. also Brown's (2001) call for 'tormenting' customers through retromarketing.

29. Prospecting is the process of finding 'the relevant pockets of knowledge from around the world' (Santos, Doz, and Williamson 2004: 35).

30. This case study is based on Kohlbacher and Mukai (2007).

31. As of 2006, Asia-Pacific headquarters were relocated to Shanghai.

32. Ichijo and Kohlbacher (2006a, 2006b, 2007) served as helpful references in researching and writing up this case study.

33. Nonaka and Katsumi's (2006) Japanese article on the development of the roadster served as a very helpful basis for researching and writing up this case study, and a preliminary version of the case study was presented at a research workshop (Nonaka, Kohlbacher, and Holden, 2006), where useful feedback was obtained.

34. Cusumano and Nobeoka (1998: 196) call the development team of the first Mazda roadster a 'guerrilla-type' product team as a special project.

35. The official English name is 'MAYEKAWA MFG. CO., LTD'; however I use the more common transcription 'Maekawa' in this book.

36. For a review of the recent literature on knowledge brokers see Vicari and Cillo (2006).

37. Note that customer focus here refers to the initiative implemented at Siemens rather than the academic concept of customer focus that was mentioned in Chapter 4.1.3.3.

38. Of course, there will also be customers who gain value from simply owning the car. But I argue that the value will be greater if it is co-created through the experience of 'lots of fun' and by actually realizing *Jinba Ittai*.

39. Ikujiro Nonaka confirmed this to me in a personal communication on 6 July 2006 in Tokyo.

40. For detailed information on the MAKE award see the homepage of Teleos and the KNOW Network: http://www.knowledgebusiness.com/.

41. I prefer to refer to topics rather than questions. Qualitative interivews are mainly led by the interviewee rather than the interviewer. Therefore, I basically only suggested topics to talk about and asked specfifc questions in response to what the interviewees were saying.

References

Acedo F. J., C. Barroso, and J. L. Galan (2006) 'The resource-based theory: Dissemination and main trends', *Strategic Management Journal*, 27(7): 621–36.

Achrol R. S. (1991) 'Evolution of the marketing organization: New forms for turbulent environments', *Journal of Marketing*, 55(4): 77–93.

Achrol R. S. (1997) 'Changes in the theory of interorganizational relations in marketing: Toward a network paradigm', *Journal of the Academy of Marketing Science*, 25(1): 56–71.

Achrol R. S., and P. Kotler (1999) 'Marketing in the network economy', *Journal of Marketing*, 63(4): 146–62.

Ahmadjian C. L. (2004) 'Inter-organizational knowledge creation: Knowledge and networks', in H. Takeuchi and I. Nonaka (eds), *Hitotsubashi on Knowledge Management* (Singapore: John Wiley & Sons (Asia)): 227–45.

Ahmadjian C. L., and J. R. Lincoln (2001) 'Keiretsu, governance and learning: Case studies in change from the Japanese automotive industry', *Organization Science*, 12(6): 683–701.

Akutsu S., and I. Nonaka (2004) 'Branding capabilities: A look at Sony's capabilities in brand knowledge creation', in H. Takeuchi and I. Nonaka (eds), *Hitotsubashi on Knowledge Management* (Singapore: John Wiley & Sons (Asia)): 287–308.

Al-Laham A., and T. L. Amburgey (2005) 'Knowledge sourcing in foreign direct investments: An empirical examination of target profiles', *Management International Review*, 45(3): 247–75.

Alvesson M., D. Kärreman, and J. Swan (2002) 'Departures from knowledge and/or management in knowledge management', *Management Communication Quarterly*, 16(2): 282–91.

AMA (2004) 'Dictionary of marketing terms: Marketing', http://www.marketing-power.com/mg-dictionary-view1862.php, accessed 25/1/2006.

AMA Task Force on the Development of Marketing Thought (1988) 'Developing, dissemination and utilizing marketing knowledge', *Journal of Marketing*, 52(4): 1–25.

Ambos B., and B. B. Schlegelmilch (2005) 'In search of global advantage', *European Business Forum*, 21 (Spring): 23–4.

Amit R., and P. J. H. Schoemaker (1993) 'Strategic assets and organizational rent', *Strategic Management Journal*, 14(1): 33–46.

Ando K.-I. (2005) *Japanese Multinationals in Europe: a Comparison of the Automobile and Pharmaceutical Industries* (Cheltenham: Edward Elgar).

Aoshima Y. (1996) 'Knowledge transfer across generations: The impact on product development performance in the automotive industry', unpublished doctoral dissertation, Massachusetts Institute of Technology (Cambridge).

Aoshima Y. (2002) 'Transfer of system knowledge across generations in new product development: Empirical observations from Japanese automobile development', *Industrial Relations*, 41(4): 605–28.

Argyris C. (1977) 'Double loop learning in organizations', *Harvard Business Review*, 55(5): 115–25.

Argyris C., and D. A. Schön (1978) *Organizational Learning* (Reading: Addison-Wesley).

Asakawa K., and M. Lehrer (2003) 'Managing local knowledge assets globally: The role of regional innovation relays', *Journal of World Business*, 38(1): 31–42.

Atuahene-Gima K. (1996) 'Market orientation and innovation', *Journal of Business Research*, 35(2): 93–103.

Baba Y., and K. Nobeoka (1998) 'Towards knowledge-based product development: The 3-D CAD model of knowledge creation', *Research Policy*, 26(6): 643–59.

Badaracco J. L. (1991) *The Knowledge Link: How Firms Compete through Strategic Alliances* (Boston: Harvard Business School Press).

Baker M. (2000) 'Creating an alliance between employees and customers', *KM Review*, 3(5): 10–11.

Baker W. E., and J. M. Sinkula (1999a) 'The synergistic effect of market orientation and learning orientation on organizational performance', *Journal of the Academy of Marketing Science*, 27(4): 411–27.

Baker W. E., and J. M. Sinkula (1999b) 'Learning orientation, market orientation, and innovation: Integrating and extending models of organizational performance', *Journal of Market Focused Management*, 4(4): 295–308.

Baker W. E., and J. M. Sinkula (2005) 'Market orientation and the new product paradox', *Journal of Product Innovation Management*, 22(6): 483–502.

Ballantyne D. (2004) 'Action research reviewed: A market-oriented approach', *European Journal of Marketing*, 38(3/4): 321–37.

Barabba V. P., and G. Zaltman (1991) *Hearing the Voice of the Market: Competitive Advantage through Creative Use of Market Information* (Boston: Harvard Business School Press).

Barney J. B. (1991) 'Firm resources and sustained competitive advantage', *Journal of Management*, 17(1): 99–120.

Barney J. B. (2001) 'Resource-based theories of competitive advantage: A ten-year retrospective on the resource-based view', *Journal of Management*, 27(6): 643–50.

Barrett M., S. Cappleman, G. Shoib, and G. Walsham (2004) 'Learning in knowledge communities: Managing technology and context', *European Management Journal*, 22(1): 1–11.

Bartlett C. A., and S. Ghoshal (2002) *Managing across Borders: the Transnational Solution* (2nd edn) (Boston: Harvard Business School Press).

Baumard P. (1999) *Tacit Knowledge in Organizations*, trans S. Wauchope (London: Sage).

Bell D. (1973) *The Coming of Post-industrial Society: a Venture in Social Forecasting* (New York: Basic Books).

Bell S. J., G. J. Whitwell, and B. A. Lukas (2002) 'Schools of thought in organizational learning', *Journal of the Academy of Marketing Science*, 30(1): 70–86.

Benner M. J., and M. L. Tushman (2003) 'Exploitation, exploration, and process management: The productivity dilemma revisited', *Academy of Management Review*, 28(2): 238–56.

Bennet R. (1998) 'Charities, organisational learning and market orientation: A suggested measure of the propensity to behave as a learning organisation', *Journal of Marketing Practice*, 4(1): 5–25.

Bennet R., and H. Gabriel (1999) 'Organisational factors and knowledge management within large marketing departments: An empirical study', *Journal of Knowledge Management*, 3(3): 212–25.

Berry L. L., L. P. Carbone, and S. H. Haeckel (2002) 'Managing the total customer experience', *MIT Sloan Management Review*, 43(3): 85–9.

Berry M. J. A., and G. S. Linoff (1999) *Mastering Data Mining: the Art and Science of Customer Relationship Management* (New York: John Wiley & Sons).

Bertels T., and C. M. Savage (1999) 'A research agenda for the knowledge era: The tough questions', *Knowledge and Process Management*, 6(4): 205–12.

Bettis R. A. (1991) 'Strategic management and the straitjacket: An editorial essay', *Organization Science*, 2(3): 315–19.

Birkinshaw J. (2001) 'Why is knowledge management so difficult?' *Business Strategy Review*, 12(1): 11–18.

Bjerre M., and D. D. Sharma (2003) 'Is marketing knowledge international? A case of key accounts', in A. Blomstermo and D. D. Sharma (eds), *Learning in the Internationalisation Process of Firms* (Cheltenham: Edward Elgar), 123–41.

Bloom P. N. (1987) *Knowledge Development in Marketing: the MSI Experience* (Lexington: Lexington Books).

Boisot M. H. (1998) *Knowledge Assets: Securing Competitive Advantage in the Information Economy* (New York: Oxford University Press).

Bresman H., J. Birkinshaw, and R. Nobel (1999) 'Knowledge transfer in international acquisitions', *Journal of International Business Studies*, 30(3): 439–62.

Brodie R. J. (2002) 'The challenge to include relational concepts', *Marketing Theory*, 2(4): 339–43.

Brown J. S., and P. Duguid (1991) 'Organizational learning and communities-of-practice: Toward a unified view of working, learning and innovation', *Organization Science*, 2(1): 40–57.

Brown J. S., and P. Duguid (2001) 'Knowledge and organizations: A social-practice perspective', *Organization Science*, 12(2): 198–213.

Brown S. (2001) 'Torment your customers (they'll love it)', *Harvard Business Review*, 79(9): 82–8.

Brown S. L., and K. M. Eisenhardt (1995) 'Product development: Past research, present findings, and future directions', *Academy of Management Review*, 20(2): 343–78.

Brown S. L., and K. M. Eisenhardt (1997) 'The art of continuous change: Linking complexity theory and time-paced evolution in relentlessly shifting organizations', *Administrative Science Quarterly*, 42(1): 1–34.

Bryman A. (2004) *Social Research Methods* (2nd edn) (New York: Oxford University Press).

Buber R., J. Gadner, and L. Richards (eds) (2004) *Applying Qualitative Methods to Marketing Management Research* (Basingstoke: Palgrave Macmillan).

Buckley P. J., K. W. Glaister, and R. Husan (2002) 'International joint ventures: Partnering skills and cross-cultural issues', *Long Range Planning*, 35: 113–34.

Buckman R. H. (2004) *Building a Knowledge-driven Organization* (New York: McGraw-Hill).

Burton-Jones A. (1999) *Knowledge Capitalism: Business, Work, and Learning in the New Economy* (New York: Oxford University Press).

Calantone R. J., S. T. Cavusgil, and Y. Zhao (2002) 'Learning orientation, firm innovation capability, and firm performance', *Industrial Marketing Management*, 31(6): 515–24.

Carbone L. P., and S. H. Haeckel (1994) 'Engineering customer experiences', *Marketing Management*, 3(3): 8–19.

Carson D., A. Gilmore, C. Perry, and K. Grønhaug (2001) *Qualitative Marketing Research* (London: Sage).

Cassell C., and G. Symon (1994) 'Qualitative research in work contexts', in C. Cassell and G. Symon (eds), *Qualitative Methods in Organizational Research: a Practical Guide* (London: Sage), 1–13.

Cavusgil S. T. (1998) 'Perspectives: Knowledge development in international marketing', *Journal of International Marketing*, 6(2): 103–12.

Cavusgil S. T., R. J. Calantone, and Y. Zhao (2003) 'Tacit knowledge transfer and firm innovation capability', *Journal of Business & Industrial Marketing*, 18(1): 6–21.

Chakravarty A. K. (2000) *Market Driven Enterprise: Product Development, Supply Chains, and Manufacturing* (Hoboken: John Wiley & Sons).

Chaston I. (1999) *Entrepreneurial Marketing* (London: Macmillan Business).

Chaston I. (2004) *Knowledge-based Marketing: the Twenty-first Century Competitive Edge* (Thousand Oaks: Sage Publications).

Chaston I., B. Badger, and E. Sadler-Smith (2000) 'Organizational learning style and competences: A comparative investigation of relationship and trans-actionally oriented small UK manufacturing firms', *European Journal of Marketing*, 34(5/6): 625–40.

Chaston I., B. Badger, and E. Sadler-Smith (2001) 'Organizational learning: An empirical assessment of process in small UK manufacturing firms', *Journal of Small Business Management*, 39(2): 139–51.

Chaston I., B. Badger, T. Mangles, and E. Sadler-Smith (2001) 'Organisational learning style, competencies and learning systems in small, UK manufacturing firms', *International Journal of Operations & Production Management*, 21(11): 1417–32.

Chaston I., B. Badger, T. Mangles, and E. Sadler-Smith (2003) 'Relationship marketing knowledge management systems and e-commerce operations in small UK accountancy practices', *Journal of Marketing Management*, 19(1–2): 109–29.

Chell E. (2004) 'Critical incident technique', in C. Cassell and G. Symon (eds), *Essential Guide to Qualitative Methods in Organizational Research* (London, Thousand Oaks, New Delhi: Sage Publications), 45–60.

Child J., D. Faulkner, and S. B. Tallman (2005) *Cooperative Strategy: Managing Alliances, Networks, and Joint Ventures* (2nd edn) (New York: Oxford University Press).

Chini T. C. (2004) *Effective Knowledge Transfer in Multinational Corporations* (Basingstoke: Palgrave Macmillan).

Choi C. J., and S. H. Lee (1997) 'A knowledge-based view of cooperative inter-organizational relationships', in P. W. Beamish and P. J. Killing (eds), *Cooperative Strategies* (San Francisco: New Lexington Press).

Choo C. W. (2003) 'Perspectives on managing knowledge in organizations', *Cataloging & Classification Quarterly*, 37(1–2): 205–20.

Choo C. W., and N. Bontis (eds) (2002a) *The Strategic Management of Intellectual Capital and Organizational Knowledge* (New York: Oxford University Press).

Choo C. W., and N. Bontis (2002b) 'Knowledge, intellectual capital, and stra-tegy: Themes and tensions', in C. W. Choo and N. Bontis (eds), *The Strategic Management of Intellectual Capital and Organizational Knowledge* (New York: Oxford University Press), 3–19.

Christensen C. M., S. Cook, and T. Hall (2005) 'Marketing malpractice: The cause and the cure', *Harvard Business Review*, 83(12): 74–83.

Ciborra C. U., and R. Andreu (2001) 'Sharing knowledge across boundaries', *Journal of Information Technology*, 16: 73–81.

Clark K. B., and T. Fujimoto (1990) 'The power of product integrity', *Harvard Business Review*, 68(6): 107–18.

Clark K. B., and T. Fujimoto (1991) *Product Development Performance: Strategy, Organization, and Management in the World Auto Industry* (Boston: Harvard Business School Press).

Clippinger J. H. (1995) 'Visualization of knowledge: Building and using intangible assets digitally', *Planning Review*, 23(6): 28–31.

Cohen D. (1998) 'Toward a knowledge context: Report on the first annual University of California Berkeley forum on knowledge and the firm', *California Management Review*, 40(3): 22–39.

Cohen W., and D. A. Levinthal (1990) 'Absorptive capacity: A new perspective on learning and innovation', *Administrative Science Quarterly*, 35(1): 128–52.

Colton S., and V. Ward (2004) 'Story as a tool to capitalize on knowledge assets', *Business Information Review*, 21(3): 172–81.

Conner K. R. (1991) 'A historical comparison of resource-based theory and five schools of thought within industrial organization economics: Do we have a new theory of the firm?' *Journal of Management*, 17(1): 121–54.

Conner K. R., and C. K. Prahalad (1996) 'A resource-based theory of the firm: Knowledge versus opportunism', *Organization Science*, 7(5): 477–501.

Constantin J. A., and R. F. Lusch (1994) *Understanding Resource Management: How to Deploy your People, Products, and Processes for Maximum Productivity* (Oxford, OH: The Planning Forum).

Contractor F. J., and P. Lorange (2002) 'The growth of alliances in the knowledge-based economy', *International Business Review*, 11(4): 485–502.

Coviello N. E., R. J. Brodie, P. J. Danaher, and W. J. Johnston (2002) 'How firms relate to their markets: An empirical examination of contemporary marketing practices', *Journal of Marketing*, 66(3): 33–46.

Cui A. S., D. A. Griffith, and S. T. Cavusgil (2005) 'The influence of competitive intensity and market dynamism on knowledge management capabilities of multinational corporation subsidiaries', *Journal of International Marketing*, 13(3): 32–53.

Curry A., and J. Curry (2000) *The Customer Marketing Method: How to Implement and Profit from Customer Relationship Management* (New York: Free Press).

Cusumano M. A., and K. Nobeoka (1998) *Thinking Beyond Lean: How Multiproject Management is Transforming Product Development at Toyota and Other Companies* (New York: Free Press).

Cusumano M. A., and A. Takeishi (1991) 'Supplier relations and management: A survey of Japanese, Japanese transplants, and U.S. auto plants', *Strategic Management Journal*, 12(8): 563–88.

Cyert R. M., P. Kumar, and J. R. Williams (1993) 'Information, market imperfections and strategy', *Strategic Management Journal*, 14 (Winter Special Issue): 47–58.

Darroch J., and R. McNaughton (2003) 'Beyond market orienation: Knowledge management and the innovativeness of New Zealand firms', *European Journal of Marketing*, 37(3/4): 572–93.

Davenport T. H. (2005) *Thinking for a Living: How to Get Better Performance and Results from Knowledge Workers* (Boston: Harvard Business School Press).

Davenport T. H., and J. G. Harris (2005) 'Automated decision making comes of age', *MIT Sloan Management Review*, 46(4): 83–9.

Davenport T. H., J. G. Harris, and A. K. Kohli (2001) 'How do they know their customers so well?' *MIT Sloan Management Review*, 42(2): 63–73.

Davenport T. H., and P. Klahr (1998) 'Managing customer support knowledge', *California Management Review*, 40(3): 195–208.

Davenport T. H., and G. J. B. Probst (eds) (2002a) *Knowledge Management Case Book: Siemens Best Practises* (2nd edn) (Weinheim: John Wiley & Sons).

Davenport T. H., and G. J. B. Probst (2002b) 'Siemens' knowledge journey', in T. H. Davenport and G. J. B. Probst (eds), *Knowledge Management Case Book: Siemens Best Practises*, 2nd edn (Weinheim: John Wiley & Sons), 10–21.

Davenport T. H., and L. Prusak (2000) *Working Knowledge: How Organizations Manage What They Know* (Boston: Harvard Business School Press).

Davis S., and J. Botkin (1994) 'The coming of knowledge-based business', *Harvard Business Review*, 72(5): 165–70.

Dawes J. (2000) 'Market orientation and profitability: Further evidence incorporating longitudinal data', *Australian Journal of Management*, 25(2): 173–200.

Day G. S. (1990) *Market Driven Strategy: Processes for Creating Value* (New York: Free Press).

Day G. S. (1994a) 'The capabilities of market-driven organizations', *Journal of Marketing*, 58(4): 37–52.

Day G. S. (1994b) 'Continuous learning about markets', *California Management Review*, 36(4): 9–31.

Day G. S. (1998) 'What does it mean to be market-driven?' *Business Strategy Review*, 9(1): 1–14.

Day G. S. (1999a) *The Market Driven Organization: Understanding, Attracting, and Keeping Valuable Customers* (New York: Free Press).

Day G. S. (1999b) 'Creating a market-driven organization', *Sloan Management Review*, 41(1): 11–22.

Day G. S. (2000) 'Managing market relationships', *Journal of the Academy of Marketing Science*, 28(1): 24–30.

Day G. S. (2003) 'Creating a superior customer-relating capability', *MIT Sloan Management Review*, 44(3): 77–82.

Day G. S., and D. B. Montgomery (1999) 'Charting new directions for marketing', *Journal of Marketing*, 63 (Special Issue): 3–13.

Day G. S., and P. Nedungadi (1994) 'Managerial representations of competitive advantage', *Journal of Marketing*, 58(2): 31–44.

Day G. S., and P. J. H. Schoemaker (2006) *Peripheral Vision: Detecting the Weak Signals that Will Make or Break Your Company* (Boston: Harvard Business School Press).

Day G. S., and R. Wensley (1988) 'Assessing advantage: A framework for diagnosing competitive strategy', *Journal of Marketing*, 52(2): 1–20.

DeLong D. W., and L. Fahey (2000) 'Diagnosing cultural barriers to knowledge management', *Academy of Management Executive*, 14(4): 113–28.

DeMarco T., and T. Lister (1999) *Peopleware: Productive Projects and Teams* (2nd edn) (New York: Dorset House Publishing).

Denrell J., N. Arvidsson, and U. Zander (2004) 'Managing knowledge in the dark: An empirical study of the reliability of capability evaluations', *Management Science*, 50(11): 1491–504.

Deshpandé R. (ed.) (1999) *Developing a Market Orientation* (Thousand Oaks: Sage).

Deshpandé R. (2001) 'From market research use to market knowledge management', in R. Deshpandé (ed.), *Using Market Knowledge* (Thousand Oaks: Sage), 1–8.

Deshpandé R., and J. U. Farley (1998) 'The market orientation construct: Correlations, culture, and comprehensiveness', *Journal of Market Focused Management*, 2(3): 237–9.

Deshpandé R., and J. U. Farley (2004) 'Organizational culture, market orientation, innovativeness, and firm performance: An international research odyssey', *International Journal of Research in Marketing*, 21(1): 3–22.

Deshpandé R., J. U. Farley, and F. E. Webster (1993) 'Corporate culture, customer orientation, and innovativeness in Japanese firms: A quadrad analysis', *Journal of Marketing*, 57(1): 23–37.

Deshpandé R., and F. E. Webster (1989) 'Organizational culture and marketing: Defining the research agenda', *Journal of Marketing*, 53(1): 3–15.

Desouza K. C. (2005) 'The new frontiers of knowledge management', in K. C. Desouza (ed.), *New Frontiers of Knowledge Management* (Basingstoke: Palgrave Macmillan), 1–10.

Desouza K. C., and Y. Awazu (2004) 'Gaining a competitive edge from your customers: Exploring three dimensions of customer knowledge', *KM Review*, 7(3): 12–15.

Desouza K. C., and Y. Awazu (2005a) 'What do they know?' *Business Strategy Review*, 16(1): 41–5.

Desouza K. C., and Y. Awazu (2005b) *Engaged Knowledge Management: Engagement with New Realities* (Basingstoke: Palgrave Macmillan).

Desouza K. C., Y. Awazu, and S. M. Jasimuddin (2005) 'Utilizing external sources of knowledge: Opening organizational channels for effective knowledge capture', *KM Review*, 8(1): 16–19.

Desouza K. C., and R. Evaristo (2003) 'Global knowledge management strategies', *European Management Journal*, 21(1): 62–7.

Dhanaraj C., M. A. Lyles, H. K. Steensma, and L. Tihanyi (2004) 'Managing tacit and explicit knowledge transfer in IJVs: The role of relational embeddedness and the impact on performance', *Journal of International Business Studies*, 35(5): 428–42.

Dickson P. R. (1992) 'Toward a general theory of competitive rationality', *Journal of Marketing*, 56(1): 69–83.

Dixon N. M. (2000) *Common Knowledge: How Companies Thrive by Sharing What they Know* (Boston, MA: Harvard Business School Press).

Dodgson M. (1993) 'Organizational learning: A review of some literatures', *Organization Studies*, 14(3): 375–94.

Doz Y., and G. Hamel (1998) *Alliance Advantage: the Art of Creating Value through Partnering* (Boston: Harvard Business School Press).

Doz Y., J. Santos, and P. Williamson (2001) *From Global to Metanational: How Companies Win in the Knowledge Economy* (Boston: Harvard Business School Press).

Doz Y., J. Santos, and P. Williamson (2003) 'The metanational: The next step in the evolution of the multinational enterprise', in J. Birkinshaw, S. Ghoshal, C. Markides, and G. Yip (eds), *The Future of the Multinational Company* (Chichester: Wiley), 154–68.

Drucker P. F. (1954) *The Practice of Management* (New York: Harper & Row).

Drucker P. (1969) *The Age of Discontinuity: Guidelines to Our Changing Society* (New York: Harper & Row).

Drucker P. (1992) 'The new society of organizations', *Harvard Business Review*, 70(5): 95–104.

Drucker P. (1993) *The Post-capitalist Society* (New York: Harper Business).

Drucker P. F. (2002) *Managing in the Next Society* (New York: Truman Talley Books, St. Martin's Griffin).

Dubin R. (1982) 'Management: Meaning, methods, and moxie', *Academy of Management Review*, 7(3): 372–9.

Dutta S., O. Narasimhan, and S. Rajiv (1999) 'Success in high-technology markets: Is marketing capability critical?' *Marketing Science*, 18(4): 547–68.

Dyck B., F. A. Starke, G. A. Mischke, and M. Mauws (2005) 'Learning to build a car: An empirical investigation of organizational learning', *Journal of Management Studies*, 42(2): 387–416.

Dyer J. H. (1994) 'Dedicated assets: Japan's manufacturing edge', *Harvard Business Review*, 72(6): 174–8.

Dyer J. H. (1996a) 'Does governance matter? *Keiretsu* alliances and asset specificity as sources of Japanese competitive advantage', *Organization Science*, 7(6): 649–66.

Dyer J. H. (1996b) 'Specialized supplier networks as a source of competitive advantage: Evidence from the auto industry', *Strategic Management Journal*, 17(4): 271–91.

Dyer J. H., and N. W. Hatch (2004) 'Using supplier networks to learn faster', *MIT Sloan Management Review*, 45(3): 57–63.

Dyer J. H., and N. W. Hatch (2006) 'Relation-specific capabilities and barriers to knowledge transfers: Creating advantage through network relationships', *Strategic Management Journal*, 27(8): 701–19.

Dyer J. H., P. Kale, and H. Singh (2001) 'How to make strategic alliances work', *MIT Sloan Management Review*, 42(4): 37–43.

Dyer J. H., P. Kale, and H. Singh (2004) 'When to ally & when to acquire', *Harvard Business Review*, 82(7/8): 109–15.

Dyer J. H., and K. Nobeoka (2000) 'Creating and managing a high-performance knowledge-sharing network: The Toyota case', *Strategic Management Journal*, 21(3): 345–67.

Dyer J. H., and W. G. Ouchi (1993) 'Japanese-style partnerships: Giving companies a competitive edge', *Sloan Management Review*, 35(1): 51–63.

Dyer J. H., and H. Singh (1998) 'The relational view: Cooperative strategy and sources of interorganizational competitive advantage', *Academy of Management Review*, 23(4): 660–79.

Earl M. J. (1997) 'Knowledge as strategy: Reflections on Skandia International and Shorko Films', in L. Prusak (ed.), *Knowledge in Organizations* (Boston: Butterworth-Heinemann), 1–15.

Easterby-Smith M., and L. Araujo (1999) 'Organizational learning: Current debates and opportunities', in M. Easterby-Smith, L. Araujo and J. Burgoyne (eds), *Organizational Learning and the Learning Organization: Developments in Theory and Practice* (London: Sage), 1–21.

Easterby-Smith M., and M. A. Lyles (2003) 'Introduction: Watersheds of organizational learning and knowledge management', in M. Easterby-Smith and M. A. Lyles (eds), *The Blackwell Handbook of Organizational Learning and Knowledge Management* (Oxford: Blackwell Publishing), 1–16.

Edvinsson L., and P. Sullivan (1996) 'Developing a model for managing intellectual capital', *European Management Journal*, 14(4): 356–64.

Eisenhardt K. M. (1989) 'Building theories from case study research', *Academy of Management Review*, 14(4): 532–50.

Eisenhardt K. M., and J. A. Martin (2000) 'Dynamic capabilities: What are they?' *Strategic Management Journal*, 21(10/11): 1105–21.

Eisenhardt K. M., and F. M. Santos (2002) 'Knowledge-based view: A new theory of strategy?' in A. M. Pettigrew, H. Thomas, and R. Whittington (eds), *Handbook of Strategy and Management* (London: Sage), 139–64.

English M. J., and W. H. Baker, Jr. (2006) *Winning the Knowledge Transfer Race: Using your Company's Knowledge Assets to Get Ahead of the Competition* (New York: McGraw-Hill).

Eriksson K., and S. Chetty (2003) 'The effect of experience and absorptive capacity on foreign market knowledge', *International Business Review*, 12(6): 673–95.

Evans P., and B. Wolf (2005) 'Collaboration rules', *Harvard Business Review*, 83(7/8): 96–104.

Fiol C. M., and M. A. Lyles (1985) 'Organizational learning', *Academy of Management Review*, 10(4): 803–13.

Flyvbjerg B. (2001) *Making Social Science Matter: Why Social Inquiry Fails and How it Can Succeed Again*, trans. S. Sampson (Cambridge: Cambridge University Press).

Flyvbjerg B. (2006a) 'Five misunderstandings about case-study research', *Qualitative Inquiry*, 12(2): 219–45.

Flyvbjerg B. (2006b) 'Making organization research matter: Power, values, and phronesis', in S. R. Clegg, C. Hardy, T. Lawrence, and W. Nord (eds), *The Sage Handbook of Organization Studies*, 2nd edn (Thousand Oaks: Sage), 370–87.

Foss N. J. (1996a) 'More critical comments on knowledge-based theories of the firm', *Organization Science*, 7(5): 519–23.

Foss N. J. (1996b) 'Knowledge-based approaches to the theory of the firm: Some critical comments', *Organization Science*, 7(5): 470–6.

Foss N. J. (ed.) (1997) *Resources, Firms, and Strategies: a Reader in the Resource-Based Perspective* (New York: Oxford University Press).

Foss N. J., and T. Pedersen (2002) 'Transferring knowledge in MNCs, the role of sources of subsidiary knowledge and organizational context', *Journal of International Management*, 8(1): 49–67.

Foss N. J., and T. Pedersen (2004) 'Organizing knowledge processes in the multinational corporation: An introduction', *Journal of International Business Studies*, 35(5): 340–9.

Fournier S., S. Dobscha, and D. G. Mick (1998) 'Preventing the premature death of relationship marketing', *Harvard Business Review*, 76(1): 42–51.

Franke N., E. von Hippel, and M. Schreier (2006) 'Finding commercially attractive user innovations: A test of lead-user theory', *Journal of Product Innovation Management*, 23(4): 301–15.

Fujimoto T. (1999) *The Evolution of a Manufacturing System at Toyota* (New York: Oxford University Press).

Furukawa I. (1999a) *Deai no 'ba' no kosoryoku – maketingu to shouhi no 'chi' no shinka* (The Imagination of the Meeting 'ba' – the Evolution of Marketing and Consumption 'Knowledge') (Tokyo: Yuhikaku).

Furukawa I. (1999b) 'Shakaiteki nettowaku to maketingu [social networks and marketing]', in I. Nonaka (ed.), *Nettowaku bijinesu no kenkyu – fureai ga tsukuru kyokan komyuniti* (Network Business Research – Empathy Communities Created through Contacts) (Tokyo: Nikkei BP), 87–154.

Garavelli C., M. Gorgoglione, and B. Scozzi (2004) 'Knowledge management strategy and organization: A perspective of analysis', *Knowledge and Process Management*, 11(4): 273–82.

Garvin D. A. (1993) 'Building a learning organization', *Harvard Business Review*, 71(4): 78–91.

Garvin D. A. (2003) *Learning in Action: Putting Organizational Learning to Work* (Boston: Harvard Business School Press).

Ghemawat P. (2005) 'Regional strategies for global leadership', *Harvard Business Review*, 83(12): 98–108.

Ghosn C. (2002) 'Saving the business without losing the company', *Harvard Business Review*, 80(1): 37–45.

Ghosn C., and P. Riès (2005) *Shift: Inside Nissan's Historic Revival*, trans. J. Cullen (New York: Currency Doubleday).

Gibbert M., M. Leibold, and G. Probst (2002) 'Five styles of customer knowledge management, and how smart companies use them to create value', *European Management Journal*, 20(5): 459–69.

Gillham B. (2000) *Case Study Research Methods* (London, New York: Continuum).

Giroux H., and J. R. Taylor (2002) 'The justification of knowledge: Tracking the translations of quality', *Management Learning*, 33(4): 497–517.

Glaser B. G., and A. Strauss (1967) *The Discovery of Grounded Theory: Strategies for Qualitative Research* (Chicago: Aldine).

Glazer R. (1991) 'Marketing in an information-intensive environment: Strategic implications of knowledge as an asset', *Journal of Marketing*, 55(4): 1–19.

Glisby M., and N. Holden (2003) 'Contextual constraints in knowledge management theory: The cultural embeddedness of Nonaka's knowledge-creating company', *Knowledge and Process Management*, 10(1): 29–36.

Glisby M., and N. Holden (2005) 'Applying knowledge management concepts to the supply chain: How a Danish firm achieved a remarkable breakthrough in Japan', *Academy of Management Executive*, 19(2): 85–9.

Gouillart F. J., and F. D. Sturdivant (1994) 'Spend a day in the life of your customers', *Harvard Business Review*, 72(1): 116–25.

Gourlay S. (2006) 'Conceptualizing knowledge creation: A critique of Nonaka's theory', *Journal of Management Studies*, 43(7): 1415–36.

Grant R. M. (1996a) 'Prospering in dynamically-competitive environments: Organizational capability as knowledge integration', *Organization Science*, 7(4): 375–87.

Grant R. M. (1996b) 'Toward a knowledge-based theory of the firm', *Strategic Management Journal*, 17 (Winter Special Issue): 109–22.

Grant R. M. (1997) 'The knowledge-based view of the firm: Implications for management practice', *Long Range Planning*, 30(3): 450–4.

Grant R. M. (2002) 'The knowledge-based view of the firm', in C. W. Choo and N. Bontis (eds), *The Strategic Management of Intellectual Capital and Organizational Knowledge* (New York: Oxford University Press), 133–48.

Grant R. M., and C. Baden-Fuller (2004) 'A knowledge accessing theory of strategic alliances', *Journal of Management Studies*, 41(1): 61–78.

Gray B., S. Matear, C. Boshoff, and P. Matheson (1998) 'Developing a better measure of market orientation', *European Journal of Marketing*, 32(9/10): 884–903.

Greenwood D. J., and M. Levin (1998) *Introduction to Action Research: Social Research for Social Change* (Thousand Oaks: Sage).

Grieves M. (2006) *Product Lifecycle Management: Driving the Next Generation of Lean Thinking* (New York: McGraw-Hill).

Griffin A. (1997) 'The effect of project and process characteristics on product development cycle time', *Journal of Marketing Research*, 34(1): 24–35.

Griffin A., and J. R. Hauser (1993) 'The voice of the customer', *Marketing Science*, 12(1): 1–27.

Griffin A., and J. R. Hauser (1996) 'Integrating R&D and marketing: A review and analysis of the literature', *Journal of Product Innovation Management*, 13(3): 191–215.

Grønhaug K. (2002) 'Is marketing knowledge useful?' *European Journal of Marketing*, 36(3): 364–403.

Grundei J. (2000) 'Organisation der Marktforschung' [Organization of market research]', *Marketing ZFP*, 4(4): 327–42.

Gruner K. E., and C. Homburg (2000) 'Does customer interaction enhance new product success?' *Journal of Business Research*, 49(1): 1–14.

Gueldenberg S., and H. Helting (2007) 'Bridging "the great divide": Nonaka's synthesis of "western" and "eastern" knowledge concepts reassessed', *Organization*, 14(1): 99–120.

Guest G., A. Bunce, and L. Johnson (2006) 'How many interviews are enough? An experiment with data saturation and variability', *Field Methods*, 18(1): 59–82.

Gulati R., and D. Kletter (2005) 'Shrinking core, expanding periphery: The relational architecture of high-performing organizations', *California Management Review*, 47(3): 77–104.

Gulati R., N. Nohria, and A. Zaheer (2000) 'Strategic networks', *Strategic Management Journal*, 21(3) (Special Issue): 203–25.

Gulati R., and J. B. Oldroyd (2005) 'The quest for customer focus', *Harvard Business Review*, 83(4): 92–101.

Gulati R., and H. Singh (1998) 'The architecture of cooperation: Managing coordination costs and appropriation concerns in strategic alliances', *Administrative Science Quarterly*, 43(4): 781–814.

Gummesson E. (2001) 'Are current research approaches in marketing leading us astray?' *Marketing Theory*, 1(1): 27–48.

Gummesson E. (2002) *Total Relationship Marketing* (Oxford: Butterworth-Heinemann).

Gummesson E. (2003a) 'All research is interpretive!' *Journal of Business & Industrial Marketing*, 18(6/7): 482–92.

Gummesson E. (2003b) 'Relationship marketing: It all happens here and now!' *Marketing Theory*, 3(1): 167–9.

Gummesson E. (2004a) 'From one-to-one to many-to-many marketing', paper presented at the Quis9 Symposium: Service Excellence in Management: Interdisciplinary Contributions (Karlstad).

Gummesson E. (2004b) 'The practical value of adequate marketing management theory', in R. Buber, J. Gadner and L. Richards (eds), *Applying Qualitative Methods to Marketing Management Research* (Basingstoke: Palgrave Macmillan), 3–31.

Gummesson E. (2005) 'Qualitative research in marketing: Road-map for a wilderness of complexity and unpredictability', *European Journal of Marketing*, 39(3/4): 309–27.

Gummesson E. (2006) 'Qualitative research in management: Addressing complexity, context and persona', *Management Decision*, 44(2): 167–79.

Gupta A. K., and V. Govindarajan (1991) 'Knowledge flows and the structure of control within multinational corporations', *Academy of Management Review*, 16(4): 768–92.

Gupta A. K., and V. Govindarajan (2000a) 'Knowledge flows within multinational corporations', *Strategic Management Journal*, 21(4): 473–96.

Gupta A. K., and V. Govindarajan (2000b) 'Knowledge management's social dimension: Lessons from nucor steel', *MIT Sloan Management Review*, 42(1): 71–80.

Hackley C. E. (1999) 'Tacit knowledge and the epistemology of expertise in strategic marketing management', *European Journal of Marketing*, 33(7/8): 720–35.

Hadley R. D., and H. I. M. Wilson (2003) 'The network model of internationalisation and experiential knowledge', *International Business Review*, 12(6): 697–717.

Haghirian P. (2003) 'Communicating knowledge within Euro-Japanese multinational corporations', unpublished doctoral dissertation, Vienna University of Economics and Business Administration (Vienna).

Hamel G. (1991) 'Competition for competence and interpartner learning within multinational corporations', *Strategic Management Journal*, 12(1): 83–103.

Hamel G., Y. L. Doz, and C. K. Prahalad (1989) 'Collaborate with your competitors – and win', *Harvard Business Review*, 67(1): 133–9.

Han J. K., N. Kim, and R. K. Srivastava (1998) 'Market orientation and organizational performance: Is innovation a missing link?' *Journal of Marketing*, 62(4): 30–45.

Hansen M. T., and N. Nohria (2004) 'How to build collaborative advantage', *MIT Sloan Management Review*, 46(1): 22–30.

Hansen M. T., N. Nohria, and T. Tierney (1999) 'What's your strategy for managing knowledge?' *Harvard Business Review*, 77(2): 106–16.

Hanvanich S., C. Dröge, and R. Calantone (2003) 'Reconceptualizing the meaning and domain of marketing knowledge', *Journal of Knowledge Management*, 7(4): 124–35.

Hartley J. (2004) 'Case study research', in C. Cassell and G. Symon (eds), *Essential Guide to Qualitative Methods in Organizational Research* (London, Thousand Oaks, New Delhi: Sage Publications), 323–33.

He Z.-L., and P.-K. Wong (2004) 'Exploration vs. exploitation: An empirical test of the ambidexterity hypothesis', *Organization Science*, 15(4): 481–94.

Hedberg B. (1981) 'How organizations learn and unlearn', in P. C. Nystrom and W. H. Starbuck (eds), *Handbook of Organizational Design: Volume 1: Adapting Organizations to their Environments* (New York: Oxford University Press), 3–27.

Hedlund G., and I. Nonaka (1993) 'Models of knowledge management in the West and Japan', in P. Lorange, B. Chakravarthy, J. Roos and A. Van de Ven (eds), *Implementing Strategic Processes: Change, Learning and Co-operation* (Oxford: Basil Blackwell), 117–44.

Heller F. (2004) 'Action research and research action: A family of methods', in C. Cassell and G. Symon (eds), *Essential Guide to Qualitative Methods in Organizational Research* (London, Thousand Oaks, New Delhi: Sage Publications), 349–60.

Henard D. H., and D. M. Szymanski (2001) 'Why some new products are more successful than others', *Journal of Marketing*, 38(3): 362–75.

Herbig P., and L. Jacobs (1996) 'Creative problem-solving styles in the USA and Japan', *International Marketing Review*, 13(2): 63–71.

Hitt M. A., and R. D. Ireland (1985) 'Corporate distinctive competence, strategy, industry and performance', *Strategic Management Journal*, 6(3): 273–93.

Hodock C. L. (1990) 'Strategies behind the winners and losers', *Journal of Business Strategy*, 11(5): 4–7.

Hoegl M., and A. Schulze (2005) 'How to support knowledge creation in new product development: An investigation of knowledge management methods', *European Management Journal*, 23(3): 263–73.

Hofer-Alfeis J., and R. van der Spek (2002) 'The knowledge strategy process – an instrument for business owners', in T. H. Davenport and G. J. B. Probst (eds), *Knowledge Management Case Book: Siemens Best Practises*, 2nd edn (Weinheim: John Wiley & Sons), 24–39.

Holbrook M. B. (1995) 'The four faces of commodification in the development of marketing knowledge', *Journal of Marketing Management*, 11(7): 641–54.

Holden N. J. (2002) *Cross-cultural Management: a Knowledge Management Perspective* (Harlow: Financial Times/Prentice Hall).

Homburg C., and C. Pflesser (2000) 'A multiple-layer model of market-oriented organizational culture: Measurement issues and performance outcomes', *Journal of Marketing Research*, 37(11): 449–62.

Homburg C., J. P. Workman Jr., and O. Jensen (2000) 'Fundamental changes in marketing organization: The movement toward a customer-focused organizational structure', *Journal of the Academy of Marketing Science*, 28(4): 459–78.

Hubbard R., R. Brodie, and S. J. Armstrong (1992) 'Knowledge development in marketing: The role of replication research', *New Zealand Journal of Business*, 14(1): 1–12.

Huber G. P. (1991) 'Organizational learning: The contributing processes and the literatures', *Organization Science*, 2(1): 88–115.

Hult G. T. M., and O. C. Ferrell (1997) 'A global learning organization structure and market information processing', *Journal of Business Research*, 40(2): 155–66.

Hult G. T. M., D. J. J. Ketchen, and S. F. Slater (2005) 'Market orientation and performance: An integration of disparate approaches', *Strategic Management Journal*, 26(12): 1173–81.

Hurley R. F., and G. T. M. Hult (1998) 'Innovation, market orientation, and organizational learning: An integration and empirical examination', *Journal of Marketing*, 62(3): 42–54.

Husserl E. (1954) *The Crisis of European Sciences and Transcendental Phenomenology*, trans. D. Carr (Evanston: Northwestern University Press).

Hustad W. (1999) 'Expectational learning in knowledge communities', *Journal of Organizational Change Management*, 12(5): 405–18.

Iansiti M. (1997) *Technological Integration: Making Critical Choices in a Turbulent World* (Boston: Harvard Business School Press).

Iansiti M., and K. B. Clark (1994) 'Integration and dynamic capability: Evidence from product development in automobiles and mainframe computers', *Industrial and Corporate Change*, 3(3): 557–605.

Iansiti M., and R. Levien (2004a) *The Keystone Advantage: What the New Dynamics of Business Ecosystems Mean for Strategy, Innovation, and Sustainability* (Boston: Harvard Business School Press).

Iansiti M., and R. Levien (2004b) 'Strategy as ecology', *Harvard Business Review*, 82(3): 68–78.

Ibata-Arens K. C. (2004) 'Alternatives to hierarchy in Japan: Business networks and civic entrepreneurship', *Asian Business & Management*, 3(3): 315–35.

Ichijo K. (2002) 'Knowledge exploitation and knowledge exploration: Two strategies for knowledge creating companies', in C. W. Choo and N. Bontis (eds), *The Strategic Management of Intellectual Capital and Organizational Knowledge* (New York: Oxford University Press), 477–83.

Ichijo K. (2004) 'From managing to enabling knowledge', in H. Takeuchi and I. Nonaka (eds), *Hitotsubashi on Knowledge Management* (Singapore: John Wiley & Sons (Asia) Pte Ltd), 125–52.

Ichijo K. (2006a) 'The strategic management of knowledge', in K. Ichijo and I. Nonaka (eds), *Knowledge Creation and Management: New Challenges for Managers* (New York: Oxford University Press), 121–45.

Ichijo K. (2006b) 'Enabling knowledge-based competence of a corporation', in K. Ichijo and I. Nonaka (eds), *Knowledge Creation and Management: New Challenges for Managers* (New York: Oxford University Press), 83–96.

Ichijo K., and F. Kohlbacher (2006a) 'TheToyota Way of strategic knowledge creation in emerging markets', paper presented at the 26th Strategic Management Society (SMS) Annual International Conference (Vienna).

Ichijo K., and F. Kohlbacher (2006b) 'Knowledge creation in Asian markets – the Toyota Way', paper presented at the 48th Annual Meeting of the Academy of International Business (Beijing).

Ichijo K., and F. Kohlbacher (2007) 'The Toyota Way of global knowledge creation: The learn local–act global strategy', *International Journal of Automotive Technology and Management*, 7(2/3): 116–34.

Ichijo K., and I. Nonaka (2006) 'Introduction: Knowledge as competitive advantage in the age of increasing globalization', in K. Ichijo and I. Nonaka (eds), *Knowledge Creation and Management: New Challenges for Managers* (New York: Oxford University Press), 3–10.

Imai K.-I., I. Nonaka, and H. Takeuchi (1985) 'Managing the new product development process: How Japanese companies learn und unlearn', in K. B. Clark, R. H. Hayes and C. Lorenz (eds), *The Uneasy Alliance: Managing the Productivity-Technology Dilemma* (Boston: Harvard Business School Press), 337–81.

Inkpen A. C. (1996) 'Creating knowledge through colloboration', *California Management Review*, 39(1): 123–40.

Inkpen A. C. (1998) 'Learning and knowledge acquisition through international strategic alliances', *The Academy of Management Executive*, 12(4): 69–80.

Inkpen A. C. (2000) 'Learning through joint ventures: A framework of knowledge acquisition', *Journal of Management Studies*, 37(7): 1019–43.

Inkpen A. C. (2002) 'Learning, knowledge management, and strategic alliances: So many studies, so many unanswered questions', in F. J. Contractor and P. Lorange (eds), *Cooperative Strategies and Alliances* (Oxford: Pergamon), 267–89.

Inkpen A. C. (2005) 'Learning through alliances: General Motors and NUMMI', *California Management Review*, 47(4): 114–36.

Inkpen A. C., and P. W. Beamish (1997) 'Knowledge, bargaining power, and the instability of international joint ventures', *Academy of Management Review*, 22(1): 177–202.

Inkpen A. C., and S. C. Currall (2004) 'The coevolution of trust, control, and learning in joint ventures', *Organization Science*, 15(5): 586–99.

Inkpen A. C., and A. Dinur (1998) 'Knowledge management processes and international joint ventures', *Organization Science*, 9(4): 454–68.

Inkpen A. C., and K. Ramaswamy (2006) *Global Strategy: Creating and Sustaining Advantage Across Borders* (New York: Oxford University Press).

Inkpen A. C., and E. W. K. Tsang (2005) 'Social capital, networks, and knowledge transfer', *Academy of Management Review*, 30(1): 146–65.

Itami H., and T. W. Roehl (1987) *Mobilizing Invisible Assets* (Cambridge, MA: Harvard University Press).

Jansen J. J. P., F. A. J. Van den Bosch, and H. W. Volberda (2005) 'Exploratory innovation, exploitative innovation, and ambidexterity: The impact of environmental and organizational antecedents', *Schmalenbach Business Review*, 57(4): 351–63.

Jaworski B. J., and A. K. Kohli (1993) 'Market orientation: Antecedents and consequences', *Journal of Marketing*, 57(3): 53–70.

Jaworski B. J., A. K. Kohli, and A. Sahay (2000) 'Market-driven versus driving markets', *Journal of the Academy of Marketing Science*, 28(1): 45–54.

Jayachandran S., K. Hewett, and P. Kaufmann (2004) 'Customer response capability in a sense-and-respond era: The role of customer knowledge process', *Journal of the Academy of Marketing Science*, 32(3): 219–33.

Johanson J., and J.-E. Vahlne (1977) 'The internationalization process of the firm – a model of knowledge development and increasing foreign market commitments', *Journal of International Business Studies*, 8(1): 23–32.

Johansson J. K., and I. Nonaka (1996) *Relentless: the Japanese Way of Marketing* (New York: HarperBusiness).

Joshi A. W., and S. Sharma (2004) 'Customer knowledge development: Antecedents and impact on new product performance', *Journal of Marketing*, 68(4): 47–59.

Kagan J. (2002) *Surprise, Uncertainty, and Mental Structures* (Cambridge, MA: Harvard University Press).

Kamath R. R., and J. K. Liker (1994) 'A second look at Japanese product development', *Harvard Business Review*, 72(6): 154–70.

Kanter R. M. (1989) *When Giants Learn to Dance* (New York: Touchstone).

Katahira H., I. Furukawa, and M. Abe (2003) *Chokokyakushugi* [Beyond customerism] (Tokyo: Toyo Keizai Shinhosha).

Kates S. M., and J. Robertson (2004) 'Adapting action research to marketing: A dialogic argument between theory and practice', *European Journal of Marketing*, 38(3/4): 418–32.

Khanna T., R. Gulati, and N. Nohria (1998) 'The dynamics of learning alliances: Competition, cooperation and relative scope', *Strategic Management Journal*, 19(3): 193–210.

Kijima T., and T. Hirai (2003) 'Vehicle development through "kansei" engineering', *SAE Technical Paper Series*, 01(0125).

Kim W. C., and R. Mauborgne (1999) 'Strategy, value innovation, and the knowledge economy', *MIT Sloan Management Review*, 40(3): 41–54.

Kim W. C., and R. Mauborgne (2004) 'Blue ocean strategy', *Harvard Business Review*, 82(10): 76–84.

Kim W. C., and R. Mauborgne (2005) *Blue Ocean Strategy: How to Create Uncontested Market Space and Make the Competition Irrelevant* (Boston: Harvard Business School Press).

Kirca A. H., S. Jayachandran, and W. O. Bearden (2004) 'Market orientation: A meta-analytic review and assessment of its antecedents and impact on performance', *Journal of Marketing*, 69(2): 24–41.

Kogut B. (1988) 'Joint ventures: Theoretical and empirical perspectives', *Strategic Management Journal*, 9(4): 319–32.

Kogut B., and U. Zander (1992) 'Knowledge of the firm, combinative capabilities, and the replication of technology', *Organization Science*, 3(3): 383–97.

Kogut B., and U. Zander (1993) 'Knowledge of the firm and the evolutionary theory of the multinational corporation', *Journal of International Business Studies*, 24(4): 625–45.

Kogut B., and U. Zander (1996) 'What firms do? Coordination, identity, and learning', *Organization Science*, 7(5): 502–18.

Kohlbacher F. (2005) 'The use of qualitative content analysis in case study research [89 paragraphs]', *Forum Qualitative Sozialforschung / Forum: Qualitative Social Research [On-line Journal]*, 7(1) Art. 21; Available at: http://www.qualitative-research.net/fqs-texte/21-06/21-06-21-e.htm (accessed 7/1/2006).

Kohlbacher F. (2006) 'Knowledge-based approaches to international marketing – in search of excellence', paper presented at the IFSAM VIIIth World Congress (Berlin).

Kohlbacher F., N. J. Holden, M. Glisby, and A. Numic (2007) 'Knowledge-based approaches to international marketing: Unleashing the power of tacit local and global knowledge to create competitive advantage', paper presented at the 49th Annual Meeting of the Academy of International Business (Indianapolis).

Kohlbacher F., and K. Mukai (2007) 'Japan's learning communities in Hewlett-Packard Consulting & Integration: Challenging one-size fits all solutions', *The Learning Organization*, 14(1): 8–20.

Kohli A. K., and B. J. Jaworski (1990) 'Market orientation: The construct, research propositions, and managerial implications', *Journal of Marketing*, 54(2): 1–18.

Kohli A. K., B. J. Jaworski, and A. Kumar (1993) 'Markor: A measure of market orientation', *Journal of Marketing Research*, 30(4): 467–78.

Kokuryo J., I. Nonaka, and M. Kataoka (2003) *Nettowaku-shakai no chishikikeiei* [Knowledge-based Management in the Network Society] (Tokyo: NTT Publishing).

Konsynski B. R., and F. W. McFarlan (1990) 'Information partnerships – shared data, shared scale', *Harvard Business Review*, 68(5): 114–20.

Kotabe M., X. Martin, and H. Domoto (2003) 'Gaining from vertical partnerships: Knowledge transfer, relationship duration, and supplier performance improvement in the U.S. and Japanese automotive industries', *Strategic Management Journal*, 24(4): 293–316.

Kotler P., D. C. Jain, and S. Maesincee (2002) *Marketing Moves: a New Approach to Profits, Growth, and Renewal* (Boston: Harvard Business School Press).

Kotler P., S. Jatusripitak, and S. Maesincee (1997) *The Marketing of Nations: a Strategic Approach to Building National Wealth* (New York: Free Press).

Kumar N., L. Scheer, and P. Kotler (2000) 'From market driven to market driving', *European Management Journal*, 18(2): 129–42.

Kusunoki K. (2004) 'Value differentiation: Organizing "know-what" for product concept innovation', in H. Takeuchi and I. Nonaka (eds), *Hitotsubashi on Knowledge Management* (Singapore: John Wiley & Sons (Asia) Pte Ltd), 153–81.

Kusunoki K., I. Nonaka, and A. Nagata (1998) 'Organizational capabilities in product development of Japanese firms: A conceptual framework and empirical findings', *Organization Science*, 9(6): 699–718.

Kyriakopoulos K., and C. Moorman (2004) 'Tradeoffs in marketing exploitation and exploration strategies: The overlooked role of market orientation', *International Journal of Research in Marketing*, 21(3): 219–40.

Lane P. J., J. E. Salk, and M. A. Lyles (2001) 'Absorptive capacity, learning, and performance in international joint ventures', *Strategic Management Journal*, 22(12): 1139–61.

Langerak F. (2003) 'An appraisal of the predictive power of market orientation', *European Management Journal*, 21(4): 447–64.

Lave J., and E. Wenger (1991) *Situated Learning: Legitimate Peripheral Participation* (New York: Cambridge University Press).

Lawer C. (2005) 'On customer knowledge co-creation and dynamic capabilities', Working Paper, Cranfield School of Management.

Lei D., M. A. Hitt, and R. Bettis (1996) 'Dynamic core competences through meta-learning and strategic context', *Journal of Management*, 22(4): 549–69.

Leibold M., G. Probst, and M. Gibbert (2002) *Strategic Management in the Knowledge Economy: New Approaches and Business Applications* (Erlangen: Publicis).

Leonard-Barton D. (1990) 'A dual methodology for case studies: Synergistic use of a longitudinal single site with replicated multiple sites', *Organization Science*, 1(3): 248–66.

Leonard-Barton D. (1991) 'Inanimate integrators: A block of wood speaks', *Design Management Journal*, 2(3): 61–7.

Leonard-Barton D. (1992) 'Core capabilities and core rigidities: A paradox in managing new product development', *Strategic Management Journal*, 13 (Special Issue): 111–25.

Leonard D. (1998) *Wellsprings of Knowledge: Building and Sustaining the Sources of Innovation* (Boston: Harvard Business School Press).

Leonard D. (2000) 'Tacit knowledge, unarticulated needs, and empathic design in new product development', in D. Morey, M. Maybury, and B. Thuraisingham (eds), *Knowledge Management: Classic and Contemporary Works* (Cambridge, MA: MIT Press), 223–37.

Leonard D. (2006) 'Market research in product development', in K. Ichijo and I. Nonaka (eds), *Knowledge Creation and Management: New Challenges for Managers* (New York: Oxford University Press), 146–57.

Leonard D., and J. F. Rayport (1997) 'Spark innovation through empathic design', *Harvard Business Review*, 75(6): 102–13.

Leonard D., and S. Sensiper (1998) 'The role of tacit knowledge in group innovation', *California Management Review*, 40(3): 112–32.

Leonard D., and W. C. Swap (2004) 'Deep smarts', *Harvard Business Review*, 82(9): 88–97.

Leonard D., and W. C. Swap (2005a) *Deep Smarts: How to Cultivate and Transfer Enduring Business Wisdom* (Boston: Harvard Business School Press).

Leonard D., and W. C. Swap (2005b) *When Sparks Fly: Harnessing the Power of Group Creativity* (Boston: Harvard Business School Press).

Leone R. P., and R. L. Schultz (1980) 'A study of marketing generalizations', *Journal of Marketing*, 44(1): 10–18.

Lesser E. L., and J. Storck (2001) 'Communities of practice and organizational performance', *IBM Systems Journal*, 40(4): 831–41.

Levinthal D. A., and J. G. March (1993) 'The myopia of learning', *Strategic Management Journal*, 14 (Winter Special Issue): 95–112.

Lewin A. Y., and H. W. Volberda (1999) 'Prolegomena on coevolution: A framework for research on strategy and new organizational forms', *Organization Science*, 10(5): 519–34.

Lewin K. (1946) 'Action research and minority problems', *Journal of Social Issues*, 2(4): 34–46.

Li T., and R. J. Calantone (1998) 'The impact of market knowledge competence on new product advantage: Conceptualization and empirical examination', *Journal of Marketing*, 62(4): 13–29.

Li T., and S. T. Cavusgil (1998) 'Decomposing the effects of market knowledge competence in new product export', *European Journal of Marketing*, 34(1/2): 57–79.

Liker J. K. (2004) *The Toyota Way: 14 Management Principles from the World's Greatest Manufacturer* (New York: McGraw-Hill).

Liker J. K., and T. Y. Choi (2004) 'Building deep supplier relationships', *Harvard Business Review*, 82(12): 104–13.

Liker J. K., and Y.-C. Yu (2000) 'Japanese automakers, U.S. suppliers and supply-chain superiority', *MIT Sloan Management Review*, 42(1): 81–93.

Lincoln J. R., C. L. Ahmadjian, and E. Mason (1998) 'Organizational learning and purchase-supply relations in Japan: Hitachi, Matsushita, and Toyota compared', *California Management Review*, 40(3): 241–64.

Lincoln J. R., M. L. Gerlach, and C. L. Ahmadjian (1996) '*Keiretsu* networks and corporate performance in Japan', *American Sociological Review*, 61(1): 67–88.

Lincoln J. R., M. L. Gerlach, and P. Takahashi (1992) '*Keiretsu* networks in the Japanese economy: A dyad analysis of intercorporate ties', *American Sociological Review*, 57(5): 561–85.

Lindkvist L. (2005) 'Knowledge communities and knowledge collectivities: A typology of knowledge work in groups', *Journal of Management Studies*, 42(6): 1189–210.

London T., and S. L. Hart (2004) 'Reinventing strategies for emerging markets: Beyond the transnational model', *Journal of International Business Studies*, 35(5): 350–70.

Long Range Planning (2004) 'Special issue on peripheral vision', *Long Range Planning*, 37(2).

Lorenzoni G., and A. Lipparini (1999) 'The leveraging of interfirm relationships as a distinctive organizational capability: A longitudinal study', *Strategic Management Journal*, 20(4): 317–38.

Lovelock C., and E. Gummesson (2004) 'Whither services marketing? In search of a new paradigm and fresh perspectives', *Journal of Service Research*, 7(1): 20–41.

Lubatkin M., J. Florin, and P. Lane (2001) 'Learning together and apart: A model of reciprocal interfirm learning', *Human Relations*, 54(10): 1353–82.

Lyles M. A. (1988) 'Learning among joint venture sophisticated firms', *Management International Review*, 28 (Special Issue): 85–98.

Lyles M. A. (1994) 'The impact of organizational learning and joint venture formation', *International Business Review*, 3(4): 459–68.

Lyles M. A., and C. R. Schwenk (1992) 'Top management, strategy and organizational knowledge structures', *Journal of Management Studies*, 29(2): 155–74.

Macharzina K., M.-J. Oesterle, and D. Brodel (2001) 'Learning in multinationals', in M. Dierkes, A. Berthoin Antal, J. Child and I. Nonaka (eds), *Handbook of Organizational Learning and Knowledge* (Oxford: Oxford University Press), 631–56.

Madhavan R., and R. Grover (1998) 'From embedded knowledge to embodied knowledge: New product development as knowledge management', *Journal of Marketing*, 62(4): 1–12.

Madhok A. (2006) 'Revisiting multinational firms' tolerance for joint ventures: A trust-based approach', *Journal of International Business Studies*, 37(1): 30–43.

Maekawa M. (2004) *Monozukuri no gokui, hitozukuri no tetsugaku* [The secret of Manufacturing, the Philosophy of Forming People] (Tokyo: Diamond).

Mahoney J. T., and J. R. Pandian (1992) 'The resource-based view within the conversation of strategic management', *Strategic Management Journal*, 13 (Summer Special Issue): 363–80.

Makino S., and A. Delios (1996) 'Local knowledge transfer and performance: Implications for alliance formation in Asia', *Journal of International Business Studies*, 27(5): 905–27.

March J. G. (1991) 'Exploration and exploitation in organizational learning', *Organization Science*, 2(1): 71–87.

Marinova D. (2004) 'Actualizing innovation effort: The impact of market knowledge diffusion in a dynamic system of competition', *Journal of Marketing*, 68(3): 13–20.

Marketing Theory (2002) 'Special issue on marketing knowledge', *Marketing Theory*, 2(4).

Martin X., and R. Salomon (2003) 'Knowledge transfer capacity and its implications for the theory of the multinational corporation', *Journal of International Business Studies*, 34(4): 356–73.

Matsuno K., J. T. Mentzer, and J. O. Rentz (2000) 'A refinement and validation of the mark or scale', *Journal of the Academy of Marketing Science*, 28(4): 527–39.

Mavin S., and S. Cavaleri (2004) 'Viewing learning organizations through a social learning lens', *The Learning Organization*, 11(2/3): 285–9.

McAdam R., and S. McCreedy (1999) 'A critical review of knowledge management models', *The Learning Organization*, 6(3): 91–100.

McIntyre S. H., and M. Sutherland (2002) 'A critical analysis into the accumulation of marketing knowledge at the level of the firm', *Marketing Theory*, 2(4): 403–18.

Menguc B., and S. Auh (2006) 'Creating a firm-level dynamic capability through capitalizing on market orientation and innovativeness', *Journal of the Academy of Marketing Science*, 34(1): 63–73.

Menon A., and P. R. Varadarajan (1992) 'A model of marketing knowledge use within firms', *Journal of Marketing*, 56(4): 53–71.

Metcalfe S. J., and A. James (2000) 'Knowledge and capabilities: A new view of the firm', in N. J. Foss and P. L. Robertson (eds), *Resources, Technology and Strategy: Explorations in the Resource-based Perspective* (London: Routledge), 31–52.

Midgley D. (2002) 'What to codify: Marketing science or marketing engineering', *Marketing Theory*, 2(4): 363–8.

Mild A., and A. Taudes (2007) 'An agent-based investigation into the new product development capability', *Computational & Mathematical Organization Theory*, 13(3): 315–31.

Miller K. D., M. Zhao, and R. Calantone (2006) 'Adding interpersonal learning and tacit knowledge to March's exploration-exploitation model', *Academy of Management Journal*, 49(4): 709–22.

Minbaeva D., T. Pedersen, I. Björkman, C. F. Fey, and H. J. Park (2003) 'MNC knowledge transfer, subsidiary absorptive capacity, and HRM', *Journal of International Business Studies*, 34(6): 586–99.

Mitchell W. J. T. (1994) *Picture Theory: Essays on Verbal and Visual Representation* (Chicago: University of Chicago Press).

Moenaert R. K., and W. E. Souder (1990) 'An information transfer model for integrating marketing and R&D personnel in new product development projects', *Journal of Product Innovation Management*, 7(2): 91–107.

Mokyr J. (2002) *The Gifts of Athena: Historical Origins of the Knowledge Economy* (Princeton: Princeton University Press).

Montoya-Weiss M., and R. Calantone (1994) 'Determinants of new product performance: A review and meta-analysis', *Journal of Product Innovation Management*, 11(5): 397–418.

Moorman C. (1995) 'Organizational market information processes: Cultural antecedents and new product outcomes', *Journal of Marketing Research*, 32(3): 318–35.

Moorman C., and A. S. Miner (1997) 'The impact of organizational memory in new product performance and creativity', *Journal of Marketing Research*, 34(1): 91–106.

Morgan R. E. (2004) 'Market-based organisational learning – theoretical reflections and conceptual insights', *Journal of Marketing Management*, 20(1/2): 67–103.

Morgan R. E., C. S. Katsikeas, and W. Appiah-Adu (1998) 'Market orientation and organizational learning capabilities', *Journal of Marketing Management*, 14(4/5): 353–81.

Morgan R. E., and C. R. Turnell (2003) 'Market-based organizational learning and market performance gains', *British Journal of Management*, 14(3): 255–74.

Mowery D. C., J. E. Oxley, and B. S. Silverman (1996) 'Strategic alliances and interfirm knowledge transfer', *Strategic Management Journal*, 17 (Winter Special Issue).

Mowery D. C., J. E. Oxley, and B. S. Silverman (2002) 'The two faces of partner-specific absorptive capacity: Learning and cospecialization in strategic alliances', in F. J. Contractor and P. Lorange (eds), *Cooperative Strategies and Alliances* (Oxford: Pergamon), 291–319.

Mudambi R. (2002) 'Knowledge management in multinational firms', *Journal of International Management*, 8(1): 1–9.

Murillo-García M., and H. Annabi (2002) 'Customer knowledge management', *Journal of the Operational Research Society*, 53(8): 875–84.

Myers J. G., W. F. Massy, and S. A. Greyser (1980) *Marketing Research and Knowledge Development: an Assessment for Marketing Management* (Englewood Cliffs: Prentice-Hall).

Nahapiet J., and S. Ghoshal (1998) 'Social capital, intellectual capital, and the organizational advantage', *Academy of Management Review*, 23(2): 242–66.

Narula R., and J. Hagedoorn (1999) 'Innovating through strategic alliances: Moving towards international partnerships and contractual agreements', *Technovation*, 19(5): 283–94.

Narver J. C., and S. F. Slater (1990) 'The effect of a market orientation on business profitability', *Journal of Marketing*, 54(4); 20–35.

Narver J. C., S. F. Slater, and D. L. MacLachlan (2004) 'Responsive and proactive market orientation and new-product success', *Journal of Product Innovation Management*, 21(5): 334–47.

Natter M., A. Mild, M. Feuerstein, G. Dorffner, and A. Taudes (2001) 'The effect of incentive schemes and organizational arrangements on the new product development process', *Management Science*, 47(8): 1029–45.

Natter M., A. Mild, A. Taudes, and C. Geberth (2004) 'Web-based knowledge management in product concept development: The DELI approach', *International Journal of Electronic Business*, 2(5): 471–9.

Nelson R. R. (1991) 'Why do firms differ, and how does it matter?' *Strategic Management Journal*, 12 (Winter Special Issue): 61–74.

Nelson R. R., and S. G. Winter (1982) *An Evolutionary Theory of Economic Change* (Cambridge: Harvard University Press).

Newell S., M. Robertson, H. Scarbrough, and J. Swan (2002) *Managing Knowledge Work* (Basingstoke: Palgrave Macmillan).

Nobeoka K. (1995) 'Inter-project learning in new product development', *Academy of Management Journal*, Best Paper Proceedings: 432–6.

Nobeoka K., and M. A. Cusumano (1997) 'Multiproject strategy and sales growth: The benefits of rapid design transfer in new product development', *Strategic Management Journal*, 18(3): 169–86.

Nonaka I. (1972) 'Organization and market: exploratory study of centralization vs decentralization', unpublished doctoral dissertation, Graduate School of Business Administration, University of California, Berkeley.

Nonaka I. (1990a) 'Redundant, overlapping organization: A Japanese approach to managing the innovation process', *California Management Review*, 32(3): 27–38.

Nonaka I. (1990b) 'Managing globalization as a self-renewing process: Experiences of Japanese MNEs', in C. A. Bartlett, Y. Doz, and G. Hedlund (eds), *Managing the Global Firm* (London: Routledge), 69–94.

Nonaka I. (1991) 'The knowledge-creating company', *Harvard Business Review*, 69(6): 96–104.

Nonaka I. (1994) 'A dynamic theory of organizational knowledge creation', *Organization Science*, 5(1): 14–34.

Nonaka I. (1996) 'The knowledge-creating company', in K. Starkey (ed.), *How Organizations Learn* (London: Internat. Thomson Business Press), 18–31.

Nonaka I. (2005) 'Managing organizational knowledge: Theoretical and methodological foundations', in K. G. Smith and M. A. Hitt (eds), *Great Minds in*

Management: the Process of Theory Development (New York: Oxford University Press), 373–93.

Nonaka I., P. Byosiere, C. C. Borucki, and N. Konno (1994) 'Organizational knowledge creation theory: A first comprehensive test', *International Business Review*, 3(4): 337–51.

Nonaka I., and A. Katsumi (2006) 'Seiko no honshitsu: Matsuda rodosuta' [The Essence of Success: Mazda roadster], *Works*, 74 (February–March): 45–9.

Nonaka I., F. Kohlbacher, and N. Holden (2006a) 'Aging and innovation: Recreating and refining high-quality tacit knowledge through phronetic leadership', paper presented at the 66th Annual Academy of Management Meeting, Critical Management Studies Research Workshop, Managing the Aging Workforce: Leadership Towards a New Weltanschauung; August 11–16, 2006, Atlanta, USA.

Nonaka I., and N. Konno (1998) 'The concept of "ba": Building a foundation for knowledge creation', *California Management Review*, 40(3): 40–54.

Nonaka I., and N. Konno (2003) *Chishiki sozo no hohoron* [Methodology of Knowledge Creation] (Tokyo: Toyo Keizai Shinhosha).

Nonaka I., N. Konno, and R. Toyama (2001) 'Emergence of "ba": A conceptual framework for the continuous and self-transcending process of knowledge creation', in I. Nonaka and T. Nishiguchi (eds), *Knowledge Emergence: Social, Technical, and Evolutionary Dimensions of Knowledge Creation* (New York: Oxford University Press), 13–29.

Nonaka I., and F. M. Nicosia (1979) 'Marketing management, its environment, and information processing: a problem in organizational design', *Journal of Business Research*, 7(4): 277–300.

Nonaka I., V. Peltokorpi, and H. Tomae (2005) 'Strategic knowledge creation: The case of Hamamatsu photonics', *International Journal of Technology Management*, 30(3/4): 248–64.

Nonaka I., and V. Peltokorpi (2006) 'Visionary knowledge management: The case of Eisai transformation', *International Journal of Learning and Intellectual Capital*, 3(2): 109–29.

Nonaka I., P. Reinmoeller, and D. Senoo (1998) 'Management focus: The "art" of knowledge: Systems to capitalize on market knowledge', *European Management Journal*, 16(6): 673–84.

Nonaka I., and H. Takeuchi (1995) *The Knowledge-creating Company: How Japanese Companies Create the Dynamics of Innovation* (New York: Oxford University Press).

Nonaka I., and R. Toyama (2002) 'A firm as a dialectical being: Towards a dynamic theory of a firm', *Industrial and Corporate Change*, 11(5): 995–1009.

Nonaka I., and R. Toyama (2003) 'The knowledge-creating theory revisited: Knowledge creation as a synthesizing process', *Knowledge Management Research & Practice*, 1(1): 2–10.

Nonaka I., and R. Toyama (2005) 'The theory of the knowledge-creating firm: Subjectivity, objectivity, and synthesis', *Industrial and Corporate Change*, 14(3): 419–36.

Nonaka I., and R. Toyama (2006a) 'Strategy as distributed phronesis', Working Paper IMIO-14, Institute of Management, Innovation & Organization, University of California, Berkeley.

Nonaka I., and R. Toyama (2006b) 'Why do firms differ? The theory of the knowledge creating firm', in K. Ichijo and I. Nonaka (eds), *Knowledge Creation* and *Management: New Challenges for Managers* (New York: Oxford University Press), 13–31.

Nonaka I., R. Toyama, and N. Konno (2000) 'SECI, ba and leadership: A unified model of dynamic knowledge creation', *Long Range Planning*, 33(1): 1–31.

Nonaka I., R. Toyama, and A. Nagata (2000) 'A firm as a knowledge-creating entity: A new perspective on the theory of the firm', *Industrial and Corporate Change*, 9(1): 1–20.

Nonaka I., G. von Krogh, and S. C. Voelpel (2006b) 'Organizational knowledge creation theory: Evolutionary paths and future advances', *Organization Studies*, 27(8): 1179–208.

Numagami T. (1998) 'The infeasibility of invariant laws in management studies: A reflective dialogue in defense of case studies', *Organization Science*, 9(1): 2–15.

O'Reilly III C. A., and J. Pfeffer (2000) *Hidden Value: How Great Companies Achieve Extraordinary Results with Ordinary People* (Boston: Harvard Business School Press).

Ogawa S. (2000) *Demando chen keiei: Ryutsugyo no shin-bijinesumoderu* [Demand Chain Management: a New Business Model for the Distribution Industry] (Tokyo: Nihon Keizai Shimbunsha).

Osland G. E., and S. T. Cavusgil (1998) 'The use of multiple-party perspectives in international joint venture research', *Management International Review*, 38(3): 191–202.

Osterloh M., and J. Frost (2002) 'Motivation and knowledge as strategic resources', in B. S. Frey and M. Osterloh (eds), *Successful Management by Motivation, Balancing Intrinsic and Extrinsic Incentives* (Berlin, Heidelberg: Springer-Verlag), 27–51.

Osterloh M., J. Frost, and B. S. Frey (2002) 'The dynamics of motivation in new organizational forms', *International Journal of the Economics of Business*, 9(1): 61–77.

Ottesen G. G., and K. Grønhaug (2002) 'Managers' understanding of theoretical concepts: The case of market orientation', *European Journal of Marketing*, 36(11/12): 1209–24.

Ottesen G. G., and K. Grønhaug (2004a) 'Exploring the dynamics of market orientation in turbulent environments: A case study', *European Journal of Marketing*, 38(8): 956–73.

Ottesen G. G., and K. Grønhaug (2004b) 'Barriers to practical use of academic marketing knowledge', *Marketing Intelligence & Planning*, 22(5): 520–30.

Parkhe A. (1993) '"Messy" research, methodological predispositions, and theory development in international joint ventures', *Academy of Management Review*, 18(2): 227–68.

Parvatiyar A., and J. N. Sheth (2000) 'The domain and conceptual foundations of relationship marketing', in J. N. Sheth and A. Parvatiyar (eds), *Handbook of Relationship Marketing* (Thousands Oaks, CA: Sage Publications), 3–38.

Patriotta G. (2003) *Organizational Knowledge in the Making: How Firms Create, Use, and Institutionalize Knowledge* (New York: Oxford University Press).

Patton E., and S. H. Appelbaum (2003) 'The case for case studies in management research', *Management Research News*, 26(5): 60–71.

Patton M. Q. (2002) *Qualitative Research and Evaluation Methods*, 3rd edn (Thousand Oaks: Sage).

Peltokorpi V., and E. Tsuyuki (2006) 'Knowledge governance in a Japanese project-based organization', *Knowledge Management Research & Practice*, 4(1): 36–45.

Penrose E. (1995) *The Theory of the Growth of the Firm*, 3rd edn (New York: Oxford University Press).

Peppers D., and M. Rogers (1997) *Enterprise One to One: Tools for Competing in the Interactive Age* (New York: Currency/Doubleday).

Peppers D., and M. Rogers (1999) *The One to One Manager: Real-world Lessons in Customer Relationship Management* (New York: Doubleday).

Peppers D., M. Rogers, and B. Dorf (1999) 'Is your company ready for one-to-one marketing?' *Harvard Business Review*, 77(1): 151–60.

Perry C. (2004) 'Realism also rules ok: Scientific paradigms and case research in marketing', in R. Buber, J. Gadner, and L. Richards (eds), *Applying Qualitative Methods to Marketing Management Research* (Basingstoke: Palgrave Macmillan), 46–57.

Perry C., and E. Gummesson (2004) 'Action research in marketing', *European Journal of Marketing*, 38(3/4): 310–20.

Peteraf M. A. (1993) 'The cornerstones of competitive advantage: A resource-based view', *Strategic Management Journal*, 14(3): 179–91.

Pfeffer J. (1992) *Managing with Power: Politics and Influence in Organizations* (Boston: Harvard Business School Press).

Pine II B. J., and J. H. Gilmore (1999) *The Experience Economy: Work is Theatre and Every Business a Stage* (Boston: Harvard Business School Press).

Pine II B. J., D. Peppers, and M. Rogers (1995) 'Do you want to keep your customers forever?' *Harvard Business Review*, 73(2): 103–14.

Pine II B. J., B. Victor, and A. C. Boynton (1993) 'Making mass customization work', *Harvard Business Review*, 71(5): 108–19.

Plaskoff J. (2003) 'Intersubjectivity and community building: Learning to learn organizationally', in M. Easterby-Smith and M. A. Lyles (eds), *The Blackwell Handbook of Organizational Learning and Knowledge Management* (Oxford: Blackwell Publishing), 161–84.

Polanyi M. (1962) *Personal Knowledge: Towards a Post-critical Philosophy* (Chicago: University of Chicago Press).

Polanyi M. (1966) *The Tacit Dimension* (New York: Doubleday).

Pollard D. (2006) 'Promoting learning transfer: Developing SME marketing knowledge in the Dnipropetrovsk Oblast, Ukraine', paper presented at the 33rd Annual Academy of International Business (AIB) UK Conference, 7–8 April 2006 (Manchester).

Porter M. E. (1980) *Competitive Strategy: Techniques for Analyzing Industries and Competitors* (New York: Free Press).

Porter M. E. (1985) *Competitive Advantage: Creating and Sustaining Superior Performance* (New York: Free Press).

Porter M. E., and M. Sakakibara (2004) 'Competition in Japan', *Journal of Economic Perspectives*, 18(1): 27–50.

Prahalad C. K., and G. Hamel (1990) 'The core competence of the corporation', *Harvard Business Review*, 68(3): 79–91.

Prahalad C. K., and V. Ramaswamy (2000) 'Co-opting customer competence', *Harvard Business Review*, 78(1): 79–87.

Prahalad C. K., and V. Ramaswamy (2003) 'The new frontier of experience innovation', *MIT Sloan Management Review*, 44(4): 12–18.

Prahalad C. K., and V. Ramaswamy (2004a) *The Future of Competition: Co-creating Unique Value with Customers* (Boston: Harvard Business School Press).

Prahalad C. K., and V. Ramaswamy (2004b) 'Co-creation experiences: The next practice in value creation', *Journal of Interactive Marketing*, 18(3): 5–14.

Probst G. J. B. (2002) 'Putting knowledge to work: Case-writing as a knowledge management and organizational learning tool', in T. H. Davenport and G. J. B. Probst (eds), *Knowledge Management Case Book: Siemens Best Practises*, 2nd edn (Weinheim: John Wiley & Sons), 312–23.

Probst G., B. Büchel, and S. Raub (1998) 'Knowledge as a strategic resource', in G. von Krogh, J. Roos and D. Kleine (eds), *Knowing in Firms: Understanding, Managing and Measuring Knowledge* (London: Sage), 240–52.

Prusak L. (2001) 'Where did knowledge management come from?' *IBM Systems Journal*, 40(4): 1002–7.

Quattrone P. (2006) 'The possibility of the testimony: A case for case study research', *Organization*, 13(1): 143–57.

Quinn J. B. (1992) *Intelligent Enterprise: a Knowledge and Service Based Paradigm for Industry* (New York: Free Press).

Rädler G. (2006) 'Toyota's strategy and initiatives in Europe: The launch of the Aygo', IMD Case Study, IMD-3-1673.

Reinmoeller P., and N. van Baardwijk (2005) 'The link between diversity and resilience', *MIT Sloan Management Review*, 46(4): 61–5.

Remenyi D., A. Money, D. Price, and F. Bannister (2002) 'The creation of knowledge through case study research', *Irish Journal of Management*, 23(2): 1–17.

Richards I., D. Foster, and R. Morgan (1998) 'Brand knowledge management: Growing brand equity', *Journal of Knowledge Management*, 2(1): 47–54.

Riesenberger J. R. (1998) 'Executive insights: Knowledge – the source of sustainable competitive advantage', *Journal of International Marketing*, 6(3): 94–107.

Rigby D. K., and D. Ledingham (2004) 'CRM done right', *Harvard Business Review*, 82(11): 118–29.

Rigby D. K., F. F. Reichheld, and P. Schefter (2002) 'Avoid the four perils of CRM', *Harvard Business Review*, 80(2): 101–9.

Rindova V. P., and C. J. Fombrun (1999) 'Constructing competitive advantage: The role of firm-constituent interactions', *Strategic Management Journal*, 20(8): 691–710.

Rossiter J. R. (2001) 'What is marketing knowledge? Stage I: Forms of marketing knowledge', *Marketing Theory*, 1(1): 9–26.

Rossiter J. R. (2002) 'The five forms of transmissible, usable marketing knowledge', *Marketing Theory*, 2(4): 369–80.

Rossiter J. R. (2005) 'What is marketing knowledge? Stage II: Evidence to establish marketing knowledge', Working Paper No. 2005–01, Marketing Discipline, School of Management and Marketing, Faculty of Commerce, University of Wollongong NSW, Australia, March.

Rubin H. J., and I. S. Rubin (1995) *Qualitative Interviewing: the Art of Hearing Data* (Thousand Oaks: Sage).

Ruekert R. W. (1992) 'Developing a market orientation: An organizational strategy perspective', *International Journal of Research in Marketing*, 9(3): 225–45.

Ryu C., Y. J. Kim, A. Chaudhury, and H. R. Rao (2005) 'Knowledge acquisition via three learning processes in enterprise information portals: Learning-by-investment, learning-by-doing, and learning-from-others', *MIS Quarterly*, 29(2): 245–78.

Saint-Onge H., and D. Wallace (2003) *Leveraging Communities of Practice for Strategic Advantage* (Boston: Butterworth-Heinemann).

Salk J. E., and B. L. Simonin (2003) 'Beyond alliances: Towards a meta-theory of collaborative learning', in M. Easterby-Smith and M. A. Lyles (eds), *The Blackwell Handbook of Organizational Learning and Knowledge Management* (Oxford: Blackwell Publishing), 253–77.

Santos-Vijande M. L., M. J. Sanzo-Pérez, L. I. Álvarez-González, and R. Vázquez-Casielles (2005) 'Organizational learning and market orientation: Interface and effects on performance', *Industrial Marketing Management*, 34(3): 187–202.

Santos J., Y. Doz, and P. Williamson (2004) 'Is your innovation process global?' *MIT Sloan Management Review*, 45(4): 31–7.

Savage C. M. (1996) *Fifth Generation Management: Co-creating Through Virtual Enterprising, Dynamic Teaming and Knowledge Networking* (Newton: Butterworth-Heinemann).

Sawhney M. (2002) 'Don't just relate – collaborate', *MIT Sloan Management Review*, 43(3): 96.

Sawhney M., and E. Prandelli (2000a) 'Beyond customer knowledge management: Customers as knowledge co-creators', in Y. Malhotra (ed.), *Knowledge Management and Virtual Organizations* (Hershey: Idea Group Publishing), 258–82.

Sawhney M., and E. Prandelli (2000b) 'Communities of creation: Managing distributed knowledge in turbulent times', *California Management Review*, 42(4): 24–54.

Sawhney M., and J. Zabin (2002) 'Managing and measuring relational equity in the network economy', *Journal of the Academy of Marketing Science*, 30(4): 313–32.

Scarbrough H., M. Robertson, and J. Swan (2005) 'Professional media and management fashion: The case of knowledge management', *Scandinavian Journal of Management*, 21(2): 197–208.

Scarbrough H., and J. Swan (2001) 'Explaining the diffusion of knowledge management: The role of fashion', *British Journal of Management*, 12(1): 3–12.

Schein E. H. (1996) 'Three cultures of management: The key to organizational learning', *Sloan Management Review*, 38(1): 9–20.

Schlegelmilch B. B., B. Ambos, and T. C. Chini (2003) 'Are you ready to learn from your offshore affiliates?' *European Business Forum*, 16 (Winter): 50–4.

Schlegelmilch B. B., and T. C. Chini (2003) 'Knowledge transfer between marketing functions in multinational companies: A conceptual model', *International Business Review*, 12(2): 215–32.

Schlegelmilch B. B., and E. Penz (2002) 'Knowledge management in marketing', *The Marketing Review*, 3(1): 5–19.

Schlegelmilch B. B., and R. Sinkovics (1998) 'Viewpoint: Marketing in the information age – can we plan for an unpredictable future?' *International Marketing Review*, 15(3): 162–70.

Schmid S., and A. Schurig (2003) 'The development of critical capabilities in foreign subsidiaries: Disentangling the role of the subsidiary's business network', *International Business Review*, 12(6): 755–82.

Schrage M. (2006) 'My customer, my co-innovator', *Strategy & Business e-news*, http://www.strategy-business.com/press/enewsarticle/enews083106 (accessed 5/9/2006).

Schreyögg G., and D. Geiger (2006) 'Developing organizational narratives: A new dimension in knowledge management', in B. Renzl, K. Matzler, and H. Hinterhuber (eds), *The Future of Knowledge Management* (Basingstoke: Palgrave Macmillan), 82–98.

Schulz M. (2001) 'The uncertain relevance of newness: Organizational learning and knowledge flows', *Academy of Management Journal*, 44(4): 661–81.

Schulz M., and L. A. Jobe (2001) 'Codification and tacitness as knowledge management strategies: An empirical exploration', *Journal of High Technology Management Research*, 12(1): 139–65.

Schulze A., and M. Hoegl (2006) 'Knowledge creation in new product development projects', *Journal of Management*, 32(2): 210–36.

Senge P. M. (1990) *The Fifth Discipline, the Art and Practice of the Learning Organization* (New York: Currency Doubleday).

Senn C. (2006) 'The executive growth factor: How Siemens invigorated its customer relationships', *Journal of Business Strategy*, 27(1): 27–34.

Senoo D., S. Akutsu, and I. Nonaka (eds) (2001) *Chishiki keiei jissenron* [On Practice: Knowledge Creation and Utilization] (Tokyo: Hakuto Shobo).

Seybold P. B. (2001) 'Get inside the lives of your customers', *Harvard Business Review*, 79(5): 80–9.

Shapiro B. P. (1988) 'What the hell is "market oriented"?' *Harvard Business Review*, 66(6): 119–25.

Shaw M. J., C. Subramaniam, G. W. Tan, and M. E. Welge (2001) 'Knowledge management and data mining for marketing', *Decision Support Systems*, 31(1): 127–37.

Sheth J. N., R. S. Sisodia, and A. Sharma (2000) 'The antecedents and consequences of customer-centric marketing', *Journal of the Academy of Marketing Science*, 28(1): 55–66.

Shimizu H., and M. Maekawa (1998) *Kyoso kara kyoso he* [From Competition to Co-creation] (Tokyo: Iwanamishoten).

Simon H. (1969) *Sciences of the Artificial* (Cambridge, MA: MIT Press).

Simonin B. L. (1999a) 'Transfer of marketing know-how in international strategic alliances: An empirical investigation of the role and antecedents of knowledge ambiguity', *Journal of International Business Studies*, 30(3): 463–90.

Simonin B. L. (1999b) 'Ambiguity and the process of knowledge transfer in strategic alliances', *Strategic Management Journal*, 20(7): 595–623.

Simonin B. L. (2004) 'An empirical investigation of the process of knowledge transfer in international strategic alliances', *Journal of International Business Studies*, 35(5): 407–27.

Singh S. (2004) *Market Orientation, Corporate Culture and Business Performance* (Aldershot: Ashgate Publishing).

Sinkovics R. R., E. Penz, and P. N. Ghauri (2005) 'Analysing textual data in international marketing research', *Qualitative Market Research*, 8(1): 9–38.

Sinkula J. M. (1994) 'Market information processing and organizational learning', *Journal of Marketing*, 58(1): 35–45.

Sinkula J. M., W. E. Baker, and T. Noordewier (1997) 'A framework for market-based organizational learning: Linking values, knowledge, and behavior', *Journal of the Academy of Marketing Science*, 25(4): 305–18.

Skyrme D. J. (1998) 'Fact or fad? Ten shifts in knowledge management', *Knowledge Management Review*, 1(3): 6–7.

Skyrme D. J. (1999) *Knowledge Networking: Creating the Collaborative Enterprise* (Oxford: Butterworth-Heinemann).

Slater S. F., and J. C. Narver (1994) 'Does competitive environment moderate the market orientation performance relationship?' *Journal of Marketing*, 58(1): 46–55.

Slater S., and J. C. Narver (1995) 'Market orientation and the learning organization', *Journal of Marketing*, 5(3): 63–74.

Slater S. F., and J. C. Narver (1998) 'Customer-led and market-oriented: Let's not confuse the two', *Strategic Management Journal*, 19(10): 1001–6.

Slater S. F., and J. C. Narver (1999) 'Market-oriented is more than being customer-led', *Strategic Management Journal*, 20(12): 1165–8.

Sobek II D. K., J. K. Liker, and A. C. Ward (1998) 'Another look at how Toyota integrates product development', *Harvard Business Review*, 76(4): 36–49.

Sobek II D. K., A. C. Ward, and J. K. Liker (1999) 'Toyota's principles of set-based concurrent engineering', *Sloan Management Review*, 40(2): 67–83.

Soekijad M., M. A. A. Huis in't Veld, and B. Enserink (2004) 'Learning and knowledge processes in inter-organizational communities of practice', *Knowledge and Process Management*, 11(1): 3–12.

Song X. M., and M. E. Parry (1993) 'How the Japanese manage the R&D-marketing interface', *Research Technology Management*, 36(4): 32–8.

Song X. M., and M. E. Parry (1997) 'The determinants of Japanese new product successes', *Journal of Marketing Research*, 34(1): 64–77.

Spear S. (2004) 'Learning to lead at Toyota', *Harvard Business Review*, 82(5): 78–86.

Spear S. (2005) 'Fixing health care from the inside, today', *Harvard Business Review*, 83(9): 78–91.

Spear S., and H. K. Bowen (1999) 'Decoding the DNA of the Toyota production system', *Harvard Business Review*, 77(5): 96–106.

Spender J. C. (1989) *Industry Recipes: an Enquiry into the Nature and Sources of Managerial Judgement* (Oxford: Basil Blackwell).

Spender J. C. (1996a) 'Organizational knowledge, learning and memory: Three concepts in search of a theory', *Journal of Organizational Change Management*, 9(1): 63–78.

Spender J. C. (1996b) 'Making knowledge the basis of a dynamic theory of the firm', *Strategic Management Journal*, 17 (Winter Special Issue): 45–62.

Spender J. C. (2003) 'Exploring uncertainty and emotion in the knowledge-based theory of the firm', *Information Technology & People*, 16(3): 266–88.

Spender J. C., and R. M. Grant (1996) 'Knowledge and the firm: Overview', *Strategic Management Journal*, 17 (Winter Special Issue): 5–9.

Srivastava R. K., T. A. Shervani, and L. Fahey (1999) 'Marketing, business processes, and shareholder value: An organizationally embedded view of marketing activities and discipline of marketing', *Journal of Marketing*, 63 (Special Issue): 168–79.

Stake R. E. (1995) *The Art of Case Study Research* (Thousand Oaks, CA: Sage Publications).

Stake R. E. (2000) 'Case studies', in N. K. Denzin and Y. S. Lincoln (eds), *Handbook of Qualitative Research* (Thousand Oaks: Sage), 435–53.

Stalk G., P. Evans, and L. E. Shulman (1992) 'Competing on capabilities: The new rules of corporate strategy', *Harvard Business Review*, 70(2): 57–69.

Steensma H. K., L. Tihanyi, M. A. Lyles, and C. Dhanaraj (2005) 'The evolving value of foreign partnerships in transitioning economies', *Academy of Management Journal*, 48(2): 213–35.

Stehr N. (1994) *Knowledge Societies* (London: Sage).

Stehr N., and R. V. Ericson (1992) 'The culture and power of knowledge in modern society', in N. Stehr and R. V. Ericson (eds), *The Culture and Power of Knowledge: Inquiries into Contemporary Societies* (Berlin: Walter de Gruyter), 3–19.

Sternberg R. J., G. B. Forsythe, J. Hedlund, J. A. Horvath, R. K. Wagner, W. M. Williams, S. A. Snook, and E. Grigorenko (2000) *Practical Intelligence in Everyday Life* (Cambridge: Cambridge University Press).

Storck J., and P. A. Hill (2000) 'Knowledge diffusion through "strategic communities"', *Sloan Management Review*, 41(2): 63–4.

Strauss A., and J. M. Corbin (1990) *Basics of Qualitative Research: Techniques and Procedures for Developing Grounded Theory* (Newbury Park: Sage).

Swan J., H. Scarbrough, and M. Robertson (2002) 'The construction of "communities of practice" in the management of innovation', *Management Learning*, 33(4): 477–96.

Swap W. C., D. Leonard, M. Shields, and L. Abrams (2001) 'Using mentoring and storytelling to transfer knowledge in the workplace', *Journal of Management Information Systems*, 18(1): 95–114.

Szulanski G. (1996) 'Exploring internal stickiness: Impediments to the transfer of best practice within the firm', *Strategic Management Journal*, 17 (Winter Special Issue): 27–43.

Szulanski G. (2003) *Sticky Knowledge: Barriers to Knowing in the Firm* (London: Sage).

Szulanski G., and R. Cappetta (2003) 'Stickiness: Conceptualizing, measuring, and predicting difficulties in the transfer of knowledge within organizations', in M. Easterby-Smith and M. A. Lyles (eds), *The Blackwell Handbook of Organizational Learning and Knowledge Management* (Oxford: Blackwell Publishing), 513–34.

Takeuchi H. (2001) 'Towards a universal management concept of knowledge', in I. Nonaka and D. J. Teece (eds), *Managing Industrial Knowledge: Creation, Transfer and Utilization* (London: Sage), 315–29.

Takeuchi H., and I. Nonaka (1986) 'The new new product development game', *Harvard Business Review*, 64(1): 137–46.

Takeuchi H., and I. Nonaka (2000) 'Reflection on knowledge management from Japan', in D. Morey, M. Maybury, and B. Thuraisingham (eds), *Knowledge Management: Classic and Contemporary Works* (Cambridge, MA: MIT Press), 183–6.

Takeuchi H., and I. Nonaka (eds) (2004) *Hitotsubashi on Knowledge Management* (Singapore: John Wiley & Sons (Asia) Pte Ltd).

Tashakkori A., and C. Teddlie (1998) *Mixed Methodology: Combining Qualitative and Quantitative Approaches* (Thousand Oaks, London, New Delhi: Sage Publications).

Taudes A., M. Trcka, and M. Lukanowicz (2002) 'Organizational learning in production networks', *Journal of Economic Behavior & Organization*, 47(2): 141–63.

Taylor C. (1993) 'To follow a rule', in C. Calhoun, E. LiPuma and M. Postone (eds), *Bourdieu: Critical Perspectives* (Cambridge: Polity Press), 45–60.

Teece D. J. (1998) 'Capturing value from knowledge assets: The new economy, markets for know-how, and intangible assets', *California Management Review*, 40(3): 55–79.

Teece D. J. (2000a) *Managing Intellectual Capital: Organizational, Strategic, and Policy Dimensions* (New York: Oxford University Press).

Teece D. J. (2000b) 'Strategies for managing knowledge assets: The role of firm structure and industrial context', *Long Range Planning*, 33(1): 35–54.

Teece D. J., G. Pisano, and A. Shuen (1997) 'Dynamic capabilities and strategic management', *Strategic Management Journal*, 18(7): 509–33.

Thomas D. C. (2002) *Essentials of International Management: a Cross-cultural Perspective* (Thousand Oaks: Sage).

Thomke S. H. (2003) *Experimentation Matters: Unlocking the Potential of New Technologies for Innovation* (Boston: Harvard Business School Press).

Thomke S. H., and E. von Hippel (2002) 'Customers as innovators: A new way to create value', *Harvard Business Review*, 80(4): 74–81.

Toffler A. (1990) *Powershift: Knowledge, Wealth, and Violence at the Edge of the 21st Century* (New York: Bantam Books).

Treacy M., and F. Wiersema (1993) 'Customer intimacy and other value disciplines', *Harvard Business Review*, 71(1): 84–93.

Trompenaars F., and P. Woolliams (2004) *Marketing Across Cultures* (Chichester: Capstone).

Tseng Y. M. (2006) 'International strategies and knowledge transfer experiences of MNC's Taiwanese subsidiaries', *Journal of the American Academy of Business*, 8(2): 120–5.

Tsoukas H. (1996) 'The firm as a distributed knowledge system: A constructionist approach', *Strategic Management Journal*, 17 (Winter Special Issue): 11–25.

Tsoukas H., and N. Mylonopoulos (2004) 'Introduction: Knowledge construction and creation in organizations', *British Journal of Management*, 15 (Supplement 1): S1–S8.

Tsoukas H., and E. Vladimirou (2001) 'What is organizational knowledge?' *Journal of Management Studies*, 38(7): 973–93.

Tsuyuki E. (2001a) 'Maekawa seisakusho: Kokyaku to no "ba no kyoso"' [Maekawa manufacturing: "ba co-creation" with customers], *Hitotsubashi Business Review*, 49(1): 132–50.

Tsuyuki E. (2001b) 'Maekawa seisakusho "kokyaku to no kyoso no bazukuri"' [Maekawa manufacturing "co-creation of ba with customers"], in D. Senoo, S. Akutsu and I. Nonaka (eds), *Chishiki keiei jissenron* [On Practice: Knowledge Creation and Utilization] (Tokyo: Hakuto Shobo), 275–320.

Tsuyuki E. (2006) 'Jusoteki "ba" no keisei to chishikisozo – maekawa seisakusho' [Formation of multilayered "ba" and knowledge creation], in I. Nonaka and R. Toyama (eds), *Chishikisozokeiei to inobeshon* [Knowledge-Creating Management and Innovation] (Tokyo: Maruzen), 50–75.

Tzokas N., and M. Saren (2004) 'Competitive advantage, knowledge and relationship marketing: Where, what and how?' *Journal of Business & Industrial Marketing*, 19(2): 124–35.

Umemoto K. (2002) 'Managing existing knowledge is not enough, knowledge management theory and practice in Japan', in C. W. Choo and N. Bontis (eds), *The Strategic Management of Intellectual Capital and Organizational Knowledge: a Collection of Readings* (New York: Oxford University Press), 463–476

Un C. A., and A. Cuervo-Cazurra (2004) 'Strategies for knowledge creation in firms', *British Journal of Management*, 15(4): 27–41.

Uncles M. (2002) 'From marketing knowledge to marketing principles', *Marketing Theory*, 2(4): 345–53.

Van den Bosch F. A. J., R. Van Wijk, and H. W. Volberda (2003) 'Absorptive capacity: Antecedents, models, and outcomes', in M. Easterby-Smith and M. A. Lyles (eds), *The Blackwell Handbook of Organizational Learning and Knowledge Management* (Oxford: Blackwell Publishing), 278–301.

Vandenbosch M., and N. Dawar (2002) 'Beyond better products: Capturing value in customer interactions', *MIT Sloan Management Review*, 43(4): 35–42.

Vandermerwe S. (2000) 'How increasing value to customers improves business results', *MIT Sloan Management Review*, 42(1): 27–37.

Vandermerwe S. (2004) 'Achieving deep customer focus', *MIT Sloan Management Review*, 45(3): 26–34.

Varadarajan P. R., and A. Menon (eds) (1993) *Enhancing Knowledge Development in Marketing: Perspectives and Viewpoints* (Chicago: American Marketing Association).

Vargo S. L., and R. F. Lusch (2004) 'Evolving to a new dominant logic for marketing', *Journal of Marketing*, 68(1): 1–17.

Vera D., and M. Crossan (2003) 'Organizational learning and knowledge management: Toward an integrative framework', in M. Easterby-Smith and M. A. Lyles (eds), *The Blackwell Handbook of Organizational Learning and Knowledge Management* (Oxford: Blackwell Publishing), 122–41.

Verona G. (1999) 'A resource-based view of product development', *Academy of Management Review*, 24(1): 132–42.

Vicari S., and P. Cillo (2006) 'Developing a brokering capacity within the firm: The enactment of market knowledge', in B. Renzl, K. Matzler, and H. Hinterhuber (eds), *The Future of Knowledge Management* (Basingstoke: Palgrave Macmillan), 184–204.

von Hippel E. (1977) 'Has a customer already developed your next product?' *Sloan Management Review*, 18(2): 63–74.

von Hippel E. (1986) 'Lead users: A source of novel product concepts', *Management Science*, 32(7): 791–806.

von Hippel E. (1988) *The Sources of Innovation* (New York: Oxford University Press).

von Hippel E. (1994) 'Sticky information and the locus of problem-solving: Implications for innovations', *Management Science*, 40(4): 429–39.

von Hippel E. (2006) *Democratizing Innovation* (Cambridge, MA: MIT Press).

von Krogh G., K. Ichijo, and I. Nonaka (2000) *Enabling Knowledge Creation: How to Unlock the Mystery of Tacit Knowledge and Release the Power of Innovation* (New York: Oxford University Press).

von Krogh G., I. Nonaka, and K. Ichijo (1997) 'Develop knowledge activists!' *European Management Journal*, 15(5): 475–83.

Waddington D. (2004) 'Participant observation', in C. Cassell and G. Symon (eds), *Essential Guide to Qualitative Methods in Organizational Research* (London, Thousand Oaks, New Delhi: Sage Publications), 154–64.

Walsh J. P., and G. R. Ungson (1991) 'Organizational memory', *The Academy of Management Review*, 16(1): 57–91.

Ward A., J. K. Liker, J. J. Cristiano, and D. K. Sobek II (1995) 'The second Toyota paradox: How delaying decisions can make better cars faster', *Sloan Management Review*, 36(3): 43–61.

Watts D. J. (2003) *Six Degrees: the Science of a Connected Age* (New York: W.W. Norton).

Wayland R. E., and P. M. Cole (1997) *Customer Connections: New Strategies for Growth* (Boston: Harvard Business School Press).

Webster F. E. (1992) 'The changing role of marketing in the corporation', *Journal of Marketing*, 56(4): 1–17.

Webster F. E. (2002) *Market-driven Management: How to Define, Develop, and Deliver Customer Value*, 2nd edn (Hoboken: John Wiley & Sons).

Webster F. E. (2005) 'A perspective on the evolution of marketing management', *Journal of Public Policy & Marketing*, 24(1): 121–6.

Weick K. E. (1991) 'The nontraditional quality of organizational learning', *Organization Science*, 2(1): 116–24.

Weick K. E. (1996) 'Drop your tools: Allegory for organizational studies', *Administrative Science Quarterly*, 41(2): 301–13.

Wenger E. (2004) 'Knowledge management as a doughnut: Shaping your knowledge strategy through communities of practice', *Ivey Business Journal*, 68(3): 1–8.

Wenger E., R. McDermott, and W. M. Snyder (2002) *Cultivating Communities of Practice* (Boston: Harvard Business School Press).

Wenger E. C., and W. M. Snyder (2000) 'Communities of practice: The organizational frontier', *Harvard Business Review*, 78(1): 139–45.

Wengler S., M. Ehret, and S. Saab (2006) 'Implementation of key account management: Who, why, and how? An exploratory study on the current implementation of key account management programs', *Industrial Marketing Management*, 35(1): 103–12.

Wernerfelt B. (1984) 'A resource-based view of the firm', *Strategic Management Journal*, 5(2): 171–80.

Wernerfelt B. (1995) 'The resource-based view of the firm: Ten years after', *Strategic Management Journal*, 16(3): 171–4.

West P. (2000) *Organisational Learning in the Automotive Sector* (London: Routledge).

Wierenga B. (2002) 'On academic marketing knowledge and marketing knowledge that marketing managers use for decision-making', *Marketing Theory*, 2(4): 355–62.

Wierenga B., and P. A. M. O. Ophuis (1997) 'Marketing decision support systems: Adoption, use, and satisfaction', *International Journal of Research in Marketing*, 14(3): 275–90.

Wierenga B., and G. van Bruggen (2000) *Marketing Management Support Systems: Principles, Tools and Implementation* (Boston: Kluwer).

Wiig K. M. (2004) *People-focused Knowledge Management: How Effective Decision Making Leads to Corporate Success* (Burlington, MA: Elsevier Butterworth-Heinemann).

Wikström S. (1996a) 'Value creation by company–consumer interaction', *Journal of Marketing Management*, 12(3): 359–74.

Wikström S. (1996b) 'The customer as co-producer', *European Journal of Marketing*, 30(4): 6–19.

Wikström S., and R. Norman (1994) *Knowledge and Value: A New Perspective on Corporate Transformation* (London: Routledge).

Wolfram Cox J., and J. Hassard (2005) 'Triangulation in organizational research: A re-presentation', *Organization*, 12(1): 109–33.

Womack J. P., and D. T. Jones (1994) 'From lean production to the lean enterprise', *Harvard Business Review*, 72(2): 93–103.

Womack J. P., and D. T. Jones (1996) 'Beyond Toyota: How to root out waste and pursue perfection', *Harvard Business Review*, 74(5): 140–58.

Womack J. P., and D. T. Jones (2003) *Lean Thinking: Banish Waste and Create Wealth in your Corporation* (revised and updated edn) (New York: Simon & Schuster).

Womack J. P., and D. T. Jones (2005a) 'Lean consumption', *Harvard Business Review*, 83(3): 58–68.

Womack J. P., and D. T. Jones (2005b) *Lean Solutions: How Companies and Customers Can Create Value and Wealth Together* (New York: Free Press).

Womack J. P., D. T. Jones, and D. Roos (1991) *The Machine that Changed the World: the Story of Lean Production* (New York: HarperPerennial).

Woodruff R. B. (1997) 'Customer value: The next source for competitive advantage', *Journal of the Academy of Marketing Science*, 25(2); 139–53.

Yanow D. (2004) 'Translating local knowledge at organizational peripheries', *British Journal of Management*, 15 (Supplement 1) S9–S25.

Yasumuro K., and D. E. Westney (2001) 'Knowledge creation and the internationalization of Japanese companies: Front-line management across borders', in I. Nonaka and T. Nishiguchi (eds), *Knowledge Emergence: Social, Technical, and Evolutionary Dimensions of Knowledge Creation* (New York: Oxford University Press), 176–93.

Yeniyurt S., S. T. Cavusgil, and G. T. M. Hult (2005) 'A global market advantage framework: The role of global market knowledge competencies', *International Business Review*, 14(1): 1–19.

Yin R. K. (1981) 'The case study crisis: Some answers', *Administrative Science Quarterly*, 26(1): 58–65.

Yin R. K. (2003a) *Case Study Research, Design and Methods*, 3rd edn (Thousand Oaks: Sage).

Yin R. K. (2003b) *Applications of Case Study Research*, 2nd edn (Thousand Oaks: Sage).

Zack M. H. (1999a) 'Managing codified knowledge', *Sloan Management Review*, 40(4): 45–58.

Zack M. H. (1999b) 'Developing a knowledge strategy', *California Management Review*, 41(3): 125–45.

Zack M. H. (2003) 'Rethinking the knowledge-based organization', *MIT Sloan Management Review*, 44(4): 67–71.

Zahra S. A., and G. George (2002) 'Absorptive capacity: A review, reconceptualization, and extension', *Academy of Management Review*, 27(2): 185–203.

Zaltman G. (1996) 'Metaphorically speaking: New technique uses multidisciplinary ideas to improve qualitative research', *Marketing Research*, 8(2): 13–20.

Zaltman G. (1997) 'Rethinking market research: Putting people back in', *Journal of Marketing Research*, 34(4): 424–31.

Zaltman G. (2003) *How Customers Think: Essential Insights into the Mind of the Market* (Boston: Harvard Business School Press).

Zhu Z. (2006) 'Nonaka meets Giddens: A critique', *Knowledge Management Research & Practice*, 4(2): 106–15.

Index